Advanced Sciences and Technologies for Security Applications

Indexed by SCOPUS

The series Advanced Sciences and Technologies for Security Applications comprises interdisciplinary research covering the theory, foundations and domain-specific topics pertaining to security. Publications within the series are peer-reviewed monographs and edited works in the areas of:

- biological and chemical threat recognition and detection (e.g., biosensors, aerosols, forensics)
- crisis and disaster management
- terrorism
- cyber security and secure information systems (e.g., encryption, optical and photonic systems)
- traditional and non-traditional security
- energy, food and resource security
- economic security and securitization (including associated infrastructures)
- transnational crime
- human security and health security
- social, political and psychological aspects of security
- recognition and identification (e.g., optical imaging, biometrics, authentication and verification)
- smart surveillance systems
- applications of theoretical frameworks and methodologies (e.g., grounded theory, complexity, network sciences, modelling and simulation)

Together, the high-quality contributions to this series provide a cross-disciplinary overview of forefront research endeavours aiming to make the world a safer place.

The editors encourage prospective authors to correspond with them in advance of submitting a manuscript. Submission of manuscripts should be made to the Editor-in-Chief or one of the Editors.

Sonny S. Patel

Community Resilience When Disaster Strikes

Security and Community Health in UK Flood Zones

Springer

Sonny S. Patel
Harvard Humanitarian Initiative
Harvard TH Chan School of Public Health
Harvard University
Boston, Massachusetts, USA

Psychological Medicine Department
King's College London
London, UK

Transcultural Conflict and Violence
Initiative
Georgia State University
Atlanta, GA, USA

ISSN 1613-5113 ISSN 2363-9466 (electronic)
Advanced Sciences and Technologies for Security Applications
ISBN 978-3-031-07991-7 ISBN 978-3-031-07992-4 (eBook)
https://doi.org/10.1007/978-3-031-07992-4

This Springer imprint is published by the registered company Springer Nature Switzerland AG
The registered company address is: Gewerbestrasse 11, 6330 Cham, Switzerland

To the inspiring community members, government officials, and public health practitioners in the counties of Kent and Surrey in the UK.

To my parents, family, friends, and especially to Aarav, Betsy, and Viaan—no matter what challenges lie ahead, know that you can accomplish it with hard work, self-belief, and a community of love. Let this body of work serve as a reminder of that.

Foreword

Disasters, of all kinds, are becoming more and more prevalent across the world. Environmental changes, globalization, and population growth—as well as an arguably tumultuous social, political, and economic climate—mean that more people than ever are experiencing disasters, whether they be geophysical, climate-related, or man-made. Many scholars have worked and continue to work on understanding the impact that surviving such a terrifying and often tragic ordeal can have. The aftermath of disasters is associated not just with death, destruction, and physical injuries but often also with psychological injuries, economic problems, fractured relationships, and a sense of grief and shock that can pervade entire communities. However, over time, we tend to see people emerge from the darkness: we see strength, growth, hope, opportunities, adaptation, and tighter bonds. It is belief in these potential positive outcomes, together with passion for empowering communities and the people within them, that drives Sonny Patel to develop a better understanding of community resilience.

This book makes an important contribution to the disaster literature, using both literature reviews and primary data to understand what is meant by community resilience and identify ways of fostering resilience in communities. 'Community resilience' is a term that gets used in different ways by different people—and so the first chapter of Patel's book explores how the term has previously been used in the literature and identifies common elements within all the different definitions. Then, focusing specifically on the severe floods in the southeast of England during 2013–2014 as a case study, Patel explores perceptions of both those affected by the floods and those at risk of flooding to investigate how communities can pull together after a traumatic incident. Importantly, the qualitative data presented in the second chapter gives a voice to those affected by the severe flooding disaster, and explores the factors *they* believe strengthened, or could strengthen, their communities. The factor analysis of questionnaire data presented in the third chapter provides yet another avenue for considering critical aspects of community resilience and ways of mitigating the potential adverse effects of a disaster on affected communities. Throughout, Patel analyzes both the literature and the primary data sensitively and thoroughly, determined to be able to define community resilience and suggest how to make communities more resilient.

This book will benefit academics, students, and practitioners alike. Patel draws upon a combination of resilience literature, quantitative data, and qualitative analysis to illuminate what is understood about community resilience by not only those working in the field but also those affected or at risk of disasters. As a result, the book is a valuable resource for those wishing to understand community resilience, untangling the many varied definitions and uses of the term, and identifying nine core elements that make a resilient community. Through these nine elements, we can understand what is needed to withstand a disaster. Despite the enormous challenges that disaster-affected communities face, this book showcases how communities find the strength and resiliency to overcome even the most difficult times.

London, UK

Dr. Samantha K. Brooks
Senior Research Associate
King's College London

Acknowledgements

It takes a village to accomplish any significant body of work like this. For me, it took multiple villages. First, thanks to the community members, government officials, and public health practitioners in the counties of Kent and Surrey in the United Kingdom who shared their knowledge, time, and experience with me. I hope I have done justice to what you shared with me and that this body of work will help in future floods. Without them, this body of work would not have been possible.

Special thanks to my parents, family, friends, colleagues, and mentors at King's College London's (KCL) Psychological Medicine Department, National Institute for Health Research Health Protection Research Unit in Emergency Preparedness and Response at KCL, KCL's Move-it Study, the Harvard Humanitarian Initiative, and NIH Fogarty Global Health Training program at Harvard University for your ongoing suggestions, edits, and support. Thanks to G. James Rubin, Richard Amlôt, Brooke Rogers, Samantha Brooks, Virginia Murray, Anna Crossley, Neil Greenberg, Rebecca Webster, Daniel Stahl, Khalida Ismail, Sridhar Venkatapuram, David Demeritt, Jonathan Samet, Heather Wipfli, Timothy B. Erickson, Valerie Dobiesz, Michael VanRooyen, Irini Albanti, Phuong Pham, Negeen Darani, Katerina Bavaro, Julian Neylan, Kristina Conroy, Annelies Kersbergen, Sriram Srinivas, Wafaie Fawzi, David Hamer, D. J. Perkins, Patricie Niyitegeka, Christina Borba, Shailey Prasad, Carol Winkler, Doug Jordan, Sofia Georgopoulou, Donela Besada, Devon Elliott, Tracy McClinton Appollis, Liz Watson, Sonya Soni, Luz Moncayo, Constance Brines, Anna Rokakis, Stacy Mittelstaedt, Chantal Laillvaux, Jeremie Dupont, Micheline Frantz, Jarrett Barrios, Andrea Davis, Lorraine Schneider, Rebecca Brenner, Elizabeth Dunn, Esmail Fadae, Stacie Constantian, Steve Wilkinson, Anaïde Nahikian, Sinead Walsh, Tara Kelly, Timiyin E-Nunu, and Daniel Song for your ongoing suggestions, edits, and support.

The research shared in this book was supported by the KCL Graduate School's International Research Award and the Psychological Medicine Department at KCL. The views expressed are those of the author and not necessarily those of any institutions.

Contents

About the Author

Sonny S. Patel, MPH, MPhil is an award-winning researcher and a former National Institutes of Health Fogarty Global Health Scholar. He is a Presidential Fellow in Transcultural Conflict and Violence Initiative at Georgia State University and a Visiting Scientist at Harvard Humanitarian Initiative, Harvard TH Chan School of Public Health. He has received his Bachelor's degree from University of California, Berkeley, his Master's degrees from University of Southern California (Public Health and Global Health Leadership) and King's College London (Psychological Medicine). He has developed protocols, programs, and training to build capacity and knowledge in communities worldwide—from subject matters in Public Health, Community Health, and Mental Health to specialized topics in Emergency Management, Community Resilience, and Disaster Risk Reduction. Patel was named a Top 40 under 40 Public Health Catalyst by the Boston Congress of Public Health. He was also bestowed with the Emerald Publishing Literati Award in 2021 for Outstanding Research Paper by the Journal of Disaster Prevention and Management. He was recognized by USAID in India as one of the top Dev Disruptors in Mental Health (2020) and by NATO peers in Ukraine as a rising star in Environmental Health and Security in Conflict Zones (2018). Patel serves as a Public Health and Global Health Executive on advisory committees and board of directors of organizations creating positive social and community impacts, such as The Yogi Berra Museum and Learning Center.

Abbreviations

CART	Communities Advancing Resilience Toolkit
CCRAM	Conjoint Community Resiliency Assessment Measure
CDR	Community Disaster Resilience
CES-D	Center for Epidemiologic Studies Depression Scale
CIDI	Composite International Diagnostic Interview
CSQ	Coping Strategies Questionnaire
DROP	Disaster Resilience of Place
EA	Environment Agency
FGs	Focus Groups
GAI	Geriatric Anxiety Inventory
GHQ-12	General Health Questionnaire
GWB	General Well-Being Scale
HSCL	Hopkins Symptom Checklist
HTQ-R	Harvard Trauma Questionnaire
IES-R	Impact of Event Scales-Revised
KCL	King's College London
NRR	National Risk Register of Civil Emergencies
NRR security outcomes	Psychological impact, social disruption, and economic harm
PHQ-4	Patient Health Questionnaire for Depression and Anxiety
PTSD	Post-Traumatic Stress Disorder
RAND	Research and Development (RAND) Corporation
SF-12	Medical Outcomes Short Form 12
STAI-T	Trait Half of State-Trait Anxiety Inventory
SWLS	Satisfaction With Life Scale
UK	United Kingdom
UN	United Nations
UNISDR	United Nations International Strategy for Disaster Reduction
USA	United States of America

List of Figures

List of Tables

Chapter 1
Concept of Community Resilience When Disaster Strikes

Abstract This Chapter introduces the topic of community resilience as it applies to disasters and the relevant background for the research studies presented in this Book. Due to the variability, the definitions of community resilience in the scientific literature provide a wide scope of potential applications in practice. However, nine common elements, derived from past studies, can further inform community resilience and its effect on the likely harmful security and health outcomes caused by disasters, such as the psychological, social, and economic impact in a community. The nine community resilience elements found were: local knowledge, community networks and relationships, communication, health, governance and leadership, resources, economic investment, preparedness, and mental outlook. This Chapter sets the stage for research studies discussed in this Book, which explores these components and community resilience further in the United Kingdom to reduce harmful security and health outcomes caused by severe floods.

Keywords Community resilience · Disasters · Disaster preparedness · Disaster risk reduction · Emergency management · Public health · Global health · Operationalization · Community practice

1.1 Background

The concept of community resilience and the background needed to explore the practical application of community resilience when disaster strikes have much scientific literature and research evidence to examine. First, the term community resilience is widely used in academic and policy literature, but the meanings of the term change from text to text. That is why further exploration is necessary if applying the concept to practice. Current definitions differ by interpretation, such as being described as a process of change [11, 18, 24], a capacity-building ability [6, 12, 21], or even an adaptation to life after a disaster [4, 7, 20]. Understanding how to build community resilience into practice continues to be desired, especially for a practical measuring tool that can help communities limit the security and health harm caused by a disaster [5, 10, 25].

S. S. Patel, *Community Resilience When Disaster Strikes*, Advanced Sciences and Technologies for Security Applications,
https://doi.org/10.1007/978-3-031-07992-4_1

Building upon a comprehensive review by Patel et al. [25], community resilience will be explored as a process that communities go through when facing and coping with the impact of a disaster in this book. Within this process, specific elements are hypothesized by Patel et al. [25] from their review findings. The authors suggest that the identified elements could increase the community's ability to have a positive trajectory in the face of uncertainty, specifically providing a reduction of psychological impact, social disruption, and economic harm to a disaster-affected community. Based on this extensive review done across several disciplines, nine core elements have been identified, which are consistently suggested as constituting community resilience as it applies to disasters. They are: local knowledge, community networks and relationships, communication, health, governance and leadership, resources, economic investment, preparedness, and mental outlook [25]. These elements could inform about the areas that need to be monitored and measured if, for instance, a scale was to be developed to measure community resilience. The research presented in this book will build upon and explore these individual elements discussed in Patel et al. [25] to shape a collective understanding and application of what community resilience is and how it can be applied in practice against potential harms caused by severe environmental hazards.

1.1.1 How to Use This Book?

Each Chapter has an abstract and can stand on its own, but the uniqueness of this book is the ability to synthesize and connect the learnings in each chapter to a broader development. It is to show how we, as researchers and practitioners, work in times of environmental uncertainty to build resilience in communities, keeping in mind the public perception associated with resilience. The overall takeaways from this book are shared in Chap. 5, based on scientific literature and research carried out among people who were directly affected by a recent flood and the general public who lived in areas defined by the Environment Agency (EA) as high-risk for severe flooding in the United Kingdom (UK). The research was carried out before the coronavirus (COVID-19) pandemic. Though much of the learnings shared in this book are specific to these UK flooding areas before the pandemic, there are many concepts and evidence shared in this Book that can apply to protecting communities against multiple disasters with additional research.

1.2 What is Community Resilience When Disaster Strikes?

Operationalizing the concept of 'community resilience' against a disaster is highly sought after by disaster-response professionals, government officials, and academics. Given growing environmental security uncertainty with climate change and population movements into large cities, weather-related disasters occur frequently and with a higher intensity than in previous years [13–15]. One way to help communities against these occurrences is to develop resilience to these events. In 2015, all United Nations Member States adopted the 2030 Agenda for Sustainable Development, which seeks a global partnership on goals to ensure peace and prosperity of the world [36]. One of 17 goals directly related to resilience when disaster strikes is "to make cities and human settlements inclusive, safe, resilient and sustainable" (goal 11). This goal and the goal of climate action (goal 13) shape the policy and development of community resilience when disaster strikes on how it can be applied and enhanced against potential security and health threats caused to communities.

Even though an expectation of a quick crisis response, governments and aid organizations cannot always help communities immediately when severe environmental hazards strike. For example, the public emergency preparedness guide provided by the Government of Canada, developed in collaboration with the Canadian Association of Chiefs of Police, Canadian Association of Fire Chiefs, Canadian Red Cross, St. John Ambulance, Salvation Army, encourages families to "be prepared to take care of yourself and your family for a minimum of 72 hours" if an emergency happens in their community because "it may take emergency workers some time to reach you" [29]. In the United States of America (USA), public awareness campaigns and websites have been created in some states, such as New Hampshire and California, to increase disaster preparedness, for example, assembling household emergency kits [2, 23]. In the recent United Nations Disaster Risk Reduction framework (Sendai Framework for Disaster Risk Reduction 2015–2030), the State is listed as chief responsibility to prevent, reduce, and recover communities affected by environmental emergencies occurring within the State [33]. However, this global disaster risk reduction policy may be difficult to implement locally and rapidly.

Even when governments and organizations can help, their resources are often stretched. In 2014, Valerie Amos, Under-Secretary-General for Humanitarian Affairs and Emergency Relief Coordinator at the United Nations, noted to the press that "the world's collective response capacity and resources are being stretched to the limit" by the number of disasters that had occurred in 2013 [35]. Similar ongoing issues of delayed response and limited resources are often seen in disaster-affected communities. For example, the intangible aspects provided by governments and organizations, such as "natural, physical, human, financial, and social resources" [27, p. 2], suggest the importance of having resources widely available and distributed in the community [3, 38, 39]. Distributive justice is another ongoing issue seen in disaster-affected communities, linked to this perceived fairness of resource allocation [22].

The ongoing attention by disaster-response professionals, government officials, and academics has focused on how best to enable and empower communities to help themselves, with a concomitant focus on factors contributing to making a community resilient to severe hazards. Accompanied by this focus is a change in the rhetoric of government, industry, and charitable organizations from discussing 'disaster vulnerability' to discussing 'disaster resilience,' which is "viewed as a more proactive and positive expression of community engagement with natural hazard reduction" [8, p. 598]. The term 'disaster' is defined by the United Nations International Strategy for Disaster Reduction (UNISDR) as:

> A serious disruption of the functioning of a community or a society involving widespread human, material, economic or environmental losses and impacts, which exceeds the ability of the affected community or society to cope using its own resources [34].

It is essential to discuss the rhetoric with disaster—that there is no such thing as a natural disaster [16, 32]. Every year environmental hazards continue to show that impacts are amplified by human decisions and social constructs related to development before the threat occurs, such as corruption in the built environment [19, 30]. In this book, the research presented used the term disaster as listed above. It is crucial to understand that the disasters seen are from severe environmental hazards striking the communities, which create a serious disruption and functioning of the security and health within these communities. It is not a natural disaster.

Other terms that the research presented in this Book had to distinguish are community and resilience. A community can be broadly defined as a constituent population such as a neighborhood, town, or city—giving a locality or specific place for people. Resilience is perceived as the ability to endure and overcome, such as a set of attributes. Individually, the terms community and resilience indicate clues to what they could mean when aligned together. Still, in reality, when a disaster strikes, community resilience suggests a process in which a constituent population allows for adaptation and growth. It is important to note that past reviews have conceptualized community resilience as being distinct from individual resilience and that a resilient community does not necessarily consist of people with strong individual resilience [8, 24–26, 37].

This book aims to illuminate further understanding of what the public perceives as community resilience when disaster strikes through research in UK communities affected by or living in high-risk areas for severe flooding. Using the term 'community resilience' might create a barrier for disaster-response professionals, government officials, and academics working directly with communities at risk of experiencing a disaster. The term is too loose and varied for people to understand what it means in practice effectively. It is largely context and experience-dependent and, therefore, may be unhelpful and add to the confusion if there is no further clarification. These issues can lead to difficulties in applying community resilience to reduce harmful security and health outcomes caused by severe environmental hazards.

1.3 Specific Case of Security and Health Harm in UK Flood Zones

This book presents further research that explores the practical meaning of what community resilience means when disaster strikes through the lens of nine core elements identified previously by Patel et al. [25]. Research projects discussed in the following chapters are presented from work carried out from 2013 to 2019—before the COVID-19 pandemic. The research work encompasses severe flooding events in parts of the UK, directly involving communities who were affected or at-risk for severe flooding.

1.3.1 Community Health Issues with Severe Flooding in the UK

Floods can cause direct damage, such as to properties, roads, and human lives, and indirect impacts, such as disruption of communications, energy, water, sewage, and public services [9]. The most common floods in England are river flooding, coastal flooding, surface water flooding, sewer flooding, and groundwater flooding [9]. Table 1.1 lists the definition for each type of flooding according to the EA [9].

These types of floods occur around the world, and according to the 2012 special report by the Intergovernmental Panel on Climate Change, the three factors of "hazard, exposure, and vulnerability" can transform a flood into an enormous disaster [17, p. 43]. The exposure to and vulnerability of social and natural systems to flooding can create a range of threats for a community and be challenging to anticipate. Based on the 2008 national assessment of flood risk by the EA, "more than 5 million people live and work in 2.4 million properties that are at risk of flooding from rivers or the sea" in England [9, p. 3]. Of these, only "55% of people living in flood risk areas knew they were at risk" in 2008–2009 [9, p. 6]. With such a proportion of the population not knowing about their community's flood risk, a community's ability to anticipate and prepare against the range of threats caused by flooding are low in building capacity. Not being risk aware has several implications for preventive and response actions to a flooding disaster.

Fortunately, most communities affected by flooding in the UK have experienced few deaths or acute injuries. For example, the devastating floods of 2007 killed only 13 people across the UK [28], while the 2013 and 2014 floods saw no deaths [31]. However, flooding can cause significant impacts consistent with harmful security and health outcomes in communities, such as social disruption, economic harm, and psychological impact.

Table 1.1 EA's definitions of common types of flooding

Flood type	Definition [*Source* Environment Agency [9], p. 7]
River flooding	When "a watercourse cannot cope with the water draining into it from the surrounding land," such as after heavy storms
Coastal flooding	"A combination of high tides and stormy conditions," which could possibly lead to a tidal surge inland and cause serious flooding
Surface water flooding	"When heavy rainfall overwhelms the drainage capacity of the local area," which could commonly occur if regular maintenance of ditches, pipes, and drainage is carried out
Sewer flooding	"When sewers are overwhelmed by heavy rainfall or when they become blocked," which mainly depends on the capacity of the sewerage system and its pipes
Groundwater flooding	"When water levels in the ground rise above surface levels," which occur at higher frequency in areas situated on top of permeable rocks or aquifers

1.3.2 UK Government Security Threat of Severe Flooding

Every two years, "the UK Government produces a classified assessment of the risks of civil emergencies facing people in the UK," and flooding is included as a risk [1, p. 11]. This assessment shapes the priorities of the UK Government and determines how flooding risk reduction strategies are conducted and evaluated. The unclassified version of this assessment, the National Risk Register of Civil Emergencies (NRR), is shared with the public to inform about the risks and the potential impacts of a civil emergency for the UK over the next five years, especially as it relates to security and health. Risks are identified by "consulting experts in government departments," and the impacts of a civil emergency are assessed according to: (1) the number of deaths directly tied to the emergency; (2) the number and type of illnesses or injuries following the onset of emergency; (3) the levels of social disruption; (4) the economic harm; and (5) the psychological impact [1, p. 55]. NRR's definitions for social disruption, economic harm, and psychological impact are presented in Table 1.2.

Suppose the collective goal of applying community resilience is to reduce potential adverse outcomes of severe environmental hazards. In that case, it needs to show an ability to do so in practice, such as limiting the psychological impact, economic harm, and social disruption, as mentioned previously. The research presented in this book relates the harmful security and health outcomes caused by severe flooding to social disruption, economic harm, and psychological impact, which the UK National Risk Register (NRR) uses to evaluate the security and health impact of likely civil emergencies [1]. As a result, these NRR evaluative outcomes (social disruption, economic harm, and psychological impact) can be used to indicate how the elements of community resilience can reduce the security and health harm caused by flooding in communities. To this end, the outcomes of psychological impact, social disruption, and economic harm will be broadly labeled as 'NRR security outcomes' throughout the rest of this Book.

Differing interpretations of community resilience could be a barrier to future development and application of programs to enhance resilience at local and regional levels. Understanding the connection between the NRR security outcomes and what

Table 1.2 NRR's definitions for social disruption, economic harm, and psychological impact

Likely impact	Definition [*Source* Cabinet Office [1], p. 55]
Social disruption	The "levels of social disruption to people's daily lives, from an inability to gain access to healthcare or schools to interruptions in supplies of essential services such as food, water and fuel, and the need for evacuation of individuals from an area"
Economic harm	"The effect on the economy overall, rather than the cost of repairs"
Psychological impact	The "psychological impact that emergencies may have, including widespread anxiety, loss of confidence or outrage that communities may experience"

flood-affected and at-risk communities constitute as community resilience will help pinpoint what communities should focus on, to become better prepared against or more resilient to flooding. A step towards this goal of applying community resilience when disaster strikes will be learning from research studies conducted in the UK on this topic. In the following Chapters, research carried out among UK members of flood-affected and at-risk communities will further discuss community resilience and its effect on a community's psychological, social, and economic impacts caused by a severe flood.

References

1. Cabinet Office (2015) National risk register for civil emergencies-2015 edition. Cabinet Office, London, UK. https://www.gov.uk/government/uploads/system/uploads/attachment_data/file/419549/20150331_2015-NRR-WA_Final.pdf
2. California Department of Public Health (2015) Be prepared California. http://www.bepreparedcalifornia.ca.gov/Pages/Home.aspx
3. Carlson J, Haffenden R, Bassett G, Buehring W, Collins III M, Folga S et al (2012) Resilience: theory and application. U.S. Department of Energy, Argonne, IL. http://www.ipd.anl.gov/anlpubs/2012/02/72218.pdf
4. Castleden M, McKee M, Murray V, Leonardi G (2011) Resilience thinking in health protection. J Public Health 33(3):369–377. https://doi.org/10.1093/pubmed/fdr027
5. Clark-Ginsberg A., McCaul B, Bremaud I, Cáceres G, Mpanje D, Patel S, Patel R (2020) Practitioner approaches to measuring community resilience: The analysis of the resilience of communities to disasters toolkit. Int J Disaster Risk Reduc, 50:101714
6. Coles E, Buckle P (2004) Developing community resilience as a foundation for effective disaster recovery. Aust J Emerg Manag 19(4):6. Retrieved from https://search.informit.com.au/documentSummary;dn=375435145094637;res=ielhss
7. Cox R, Perry K (2011) Like a fish out of water: reconsidering disaster recovery and the role of place and social capital in community disaster resilience. Am J Commun Psychol 48(3–4):395–411. https://doi.org/10.1007/s10464-011-9427-0
8. Cutter SL, Barnes L, Berry M, Burton C, Evans E, Tate E et al (2008) A place-based model for understanding community resilience to natural disasters. Glob Environ Chang 18(4):598–606. https://doi.org/10.1016/j.gloenvcha.2008.07.013
9. Environment Agency (2009) Flooding in England: a national assessment of flood risk. Environment Agency, Bristol, UK. https://www.gov.uk/government/uploads/system/uploads/attachment_data/file/292928/geho0609bqds-e-e.pdf
10. Fazey I, Carmen E, Chapin FS, Ross H, Rao-Williams J, Lyon C et al (2018) Community resilience for a 1.5°C world. Curr Opin Environ Sustain 31:30–40. https://doi.org/10.1016/j.cosust.2017.12.006
11. Fazey I, Carmen E, Ross H, Rao-Williams J, Hodgson A, Searle BA et al (2021) Social dynamics of community resilience building in the face of climate change: the case of three Scottish communities. Sustain Sci 16(5):1731–1747. https://doi.org/10.1007/s11625-021-00950-x
12. Gibson CA (2010) An integrated approach to managing disruption-related risk: life and death in a model community. J Bus Contin Emerg Plan 4(3):246–261. Retrieved from http://www.ingentaconnect.com/content/hsp/jbcep/2010/00000004/00000003/art00006
13. Guha-Sapir D, Hoyois P, Below R (2013) Annual disaster statistical review 2012: the numbers and trends. Centre for Research on the Epidemiology of Disasters, Brussels, Belgium. http://cred.be/sites/default/files/ADSR_2012.pdf

14. Guha-Sapir D, Hoyois P, Below R (2015) Annual disaster statistical review 2014: the numbers and trends. Centre for Research on the Epidemiology of Disasters, Brussels, Belgium. http://cred.be/sites/default/files/ADSR_2014.pdf
15. Guha-Sapir D, Vos F, Below R, Ponserre S (2012) Annual disaster statistical review 2011: the numbers and trends. Centre for Research on the Epidemiology of Disasters, Brussels, Belgium. http://cred.be/sites/default/files/ADSR_2011.pdf
16. Hartman CW, Squires GD (2006) There is no such thing as a natural disaster: race, class, and Hurricane Katrina. Routledge, New York
17. Intergovernmental Panel on Climate Change (2012) Managing the risks of extreme events and disasters to advance climate change adaptation. Cambridge University Press, Cambridge, UK. http://www.ipcc-wg2.gov/SREX/images/uploads/SREX-All_FINAL.pdf
18. Lemyre L, Clément M, Corneil W, Craig L, Boutette P, Tyshenko M et al (2005) A psychosocial risk assessment and management framework to enhance response to CBRN terrorism threats and attacks. Biosecur Bioterror 3(4):316–330. https://doi.org/10.1089/bsp.2005.3.316
19. Loosemore M, Sanderson D, Patel SS, Greenwald K, Sharma A, Patel RB (2021) The role of construction corruption in worsening the human impact of natural disasters: a systematic literature review. Paper presented at the Proceedings of the 37th Annual ARCOM Conference, UK
20. Luthar SS (2015) Resilience in development: a synthesis of research across five decades. Dev Psychopathol, pp 739–795
21. Manyena SB (2006) The concept of resilience revisited. Disasters 30(4):434–450. https://doi.org/10.1111/j.0361-3666.2006.00331.x
22. Mayer DM (2007) Distributive justice. In: Baumeister RF, Vohs KD (eds) Encyclopedia of social psychology. SAGE Publications Ltd., Thousand Oaks, California, pp 260–261
23. New Hampshire Government (2015) Emergency preparedness resources and tools to help you take action and be safe! http://www.readynh.gov/index.htm
24. Norris FH, Stevens SP, Pfefferbaum B, Wyche KF, Pfefferbaum RL (2008) Community resilience as a metaphor, theory, set of capacities, and strategy for disaster readiness. Am J Community Psychol 41(1–2):127–150. https://doi.org/10.1007/s10464-007-9156-6
25. Patel SS, Rogers MB, Amlôt R, Rubin GJ (2017) What do we mean by 'community resilience'? A systematic literature review of how it is defined in the literature. PLoS Curr 9
26. Paton D (2008) Community resilience: integrating individual, community and societal perspectives. In Gow K, Paton D (eds) The phoenix of natural disasters: community resilience. Nova Science, New York, pp 13–21. https://books.google.com/books?id=FEj7-FvB8xgC&dq=paton+community+resilience+2008&lr=&source=gbs_navlinks_s
27. Pfefferbaum RL, Pfefferbaum B, Van Horn RL (2011) Communities advancing resilience toolkit (CART): the CART integrated system. Oklahoma City, OK. https://www.oumedicine.com/docs/ad-psychiatry-workfiles/cart_online-final_042012.pdf?sfvrsn=2
28. Pitt M (2008) The pitt review: learning lessons from the 2007 floods. Cabinet Office, London, UK. http://webarchive.nationalarchives.gov.uk/20100807034701/http://archive.cabinetoffice.gov.uk/pittreview/_/media/assets/www.cabinetoffice.gov.uk/flooding_review/pitt_review_full%20pdf.pdf
29. Public Safety Canada (2022) Your emergency preparedness guide-72 hours: is your family prepared? https://www.getprepared.gc.ca/cnt/rsrcs/pblctns/yprprdnssgd/index-en.aspx
30. Sanderson D, Patel SS, Loosemore M, Sharma A, Greenwald K, Patel RB (2021) A review of peer-reviewed published research on corruption and disasters in the built environment. Disasters. https://doi.org/10.1111/disa.12500
31. Slingo J, Belcher S, Scaide A, McCarthy M, Saulter A, McBeath K et al (2014) The recent storms and floods in the UK. Met Office, Exeter, UK. http://www.metoffice.gov.uk/media/pdf/n/i/Recent_Storms_Briefing_Final_07023.pdf
32. Smith N (2006) There's no such thing as a natural disaster. Understanding Katrina: perspectives from the social sciences. https://items.ssrc.org/understanding-katrina/theres-no-such-thing-as-a-natural-disaster/

33. UNDRR (2015) Sendai framework for disaster risk reduction 2015–2030. United Nations, Sendai, Japan. https://www.undrr.org/publication/sendai-framework-disaster-risk-reduction-2015-2030
34. UNISDR (2009) UNISDR terminology on disaster risk reduction. http://www.unisdr.org/we/inform/terminology
35. United Nations (2014) Multiple emergencies pose great test of the global humanitarian system, UN official says [Press release]. https://www.un.org/apps/news//story.asp?NewsID=46862&Cr=humanitarian&Cr1=#.Uyh-J1yE71o
36. United Nations (2015) The 17 goals. https://sdgs.un.org/goals
37. Walsh F (2007) Traumatic loss and major disasters: strengthening family and community resilience. Fam Process 46(2):207–227
38. White RK, Edwards WC, Farrar A, Plodinec M (2015) A practical approach to building resilience in America's communities. Am Behav Sci 59(2):200–219. https://doi.org/10.1177/0002764214550296
39. Zakour MJ, Gillespie DF (2013) Community disaster vulnerability: theory, research, and practice. Springer, New York, NY. https://www.springer.com/gp/book/9781461457367

Chapter 2
What Do Flood-Affected Communities Say About Community Resilience? A Qualitative Study

Abstract This Chapter presents the critical learnings and application of community resilience from flood-affected communities in the United Kingdom (UK). Focus groups and interviews (N = 77) with people from Kent, UK, and Surrey, UK, about their 2013–2014 flooding experiences provided insights into the resilience of these communities and the nine themes discussed in the previous chapter. The reported associations between the outcomes of psychological impact, social disruption, and economic harm and the elements of community resilience were anecdotally mentioned by the majority of the focus group participants. Based on their personal experience with flooding, participants confirmed that if the activities related to the elements of community resilience were consistently applied in their community before the flood, then they would have felt less or none of the flooding impacts. The qualitative findings from the focus group discussions supported a taxonomy of the community resilience elements as being relevant and having the potential to help communities against flooding.

Keywords Community resilience · Flooding · Public health · Global health · Psychological impact · Social disruption · Economic harm · Focus groups · Interviews · Flood-affected communities · Qualitative research

2.1 Background

The broad elements of community resilience were reviewed and discussed in their importance against disasters in the previous Chap. 1. They are: local knowledge, community networks and relationships, communication, health, governance and leadership, resources, economic investment, preparedness, and mental outlook [37]. Many aspects of community resilience could potentially reduce psychological impact, social disruption, and economic harm (NRR security outcomes) [9]. However, these elements should not automatically be applied without further investigation, as they could be not useful in practice, especially in protecting against the likely harm caused by a severe flood. Gathering further information from community

S. S. Patel, *Community Resilience When Disaster Strikes*, Advanced Sciences and Technologies for Security Applications,
https://doi.org/10.1007/978-3-031-07992-4_2

members about the application and real-life examples of community resilience will help with a practical taxonomy to reduce the damage caused by floods.

For example, inequalities can occur with the redevelopment and distribution of resources in communities. Wheater [56] and Wilby et al. [57] discussed how changes in the land-uses, such as redeveloping resources and areas in the communities, can increase the adverse consequences of floods. Also, the oversite and rules to the changes of land uses have been decentralized and shifted from a central government system to local authority, where more significant social inequalities can occur due to the uneven distribution of proper resources between authorities [2, 54].

2.1.1 Applying the Elements of Community Resilience to Flooding

Understanding the list of elements and hearing their connections with the NRR security outcomes can help identify the critical issues for a community and its members when facing a flood. However, addressing the problems related to flooding may vary between communities and seasons. But, determining whether the elements of community resilience can be found as being used in flood-affected communities will begin to help answer if the concept is understood well enough to be applied to communities in the United Kingdom (UK). For example, a better understanding of community resilience can be found in the range of issues that communities face with flooding described to be reduced by the elements. A key question to ask is, If the elements do have an effect in helping the communities against flooding, then in what way does this occur to reduce harmful security and health outcomes, and are there any elements missing from the list generated from the literature?

To provide further clarity on this question, qualitative evidence is needed from interviews and focus groups of people who have been affected by flooding in the UK. These flood-affected individuals can provide a better conceptualization of how the broad elements of community resilience found by Patel et al. [37], specifically relate or do not relate to the actual flooding experiences. Additionally, the first-hand qualitative research with flood-affected communities will provide valuable details for each element that would not be captured well through a quantitative survey.

2.1.2 The 2013–2014 Floods in Southeast England

Many communities in the UK are at risk from the dangers of extreme weather. Recently, communities were disrupted by severe weather during the winter season of 2013 to 2014. In a 2014 government report, the UK Meteorological Office and Centre for Ecology and Hydrology [46, p. 2] stated that the winter season of 2013 to 2014 has affected the UK with an "exceptional run of winter storms, culminating in serious coastal damage and widespread, persistent flooding." The winter season of 2013 to 2014 was "one of, if not the most, exceptional periods of winter rainfall in at least 248 years" for England and Wales [46, p. 2]. The Meteorological Office report suggested that "there is an increasing body of evidence that extreme daily rainfall rates are becoming more intense" at a national level [46, p. 3].

Two areas where intense rainfall rates affected communities in the winter during the season 2013 to 2014 were the areas of Kent and Surrey. For example, in a 2014 report by the Maidstone Borough Council [15, p. 2], the council stated that their borough had "over 300 residential and 60 business properties flooded at Christmas [in 2013] and at the start of the year [in 2014]." Situated southeast and southwest (respectively) of London, areas within Kent and Surrey have been known historically to flood to a similar effect described by the Maidstone Borough Council. Several flood defenses have been established in both areas to address the flooding, and ongoing discussions between the UK Environmental Agency (EA) and several communities in the regions have created more defenses (Kent County Council's Flood Risk Manager, personal communication, March 6, 2015). In Kent, the attention has been focused on coastal flooding and the River Medway flowing from Tonbridge to Yalding, which eventually leads into Maidstone, the capital of Kent County.

In Surrey, the attention has been focused on the River Thames flowing from Windsor to Kingston, which eventually leads into London. Based on the 2014 government report, the number of consecutive days with high flow (defined as greater than 275 $m^3 s^{-1}$) seen at the River Thames at Kingston during the winter season of 2013 to 2014 was higher than "the most extensive flooding in England and Wales during the twentieth century" in 1947 [46, p. 10]. Smaller rivers and streams, such as the Bourne in Surrey and the Nail Bourne and the Little Stour in Kent, tend to be overlooked for defenses and even necessary maintenance to mitigate possible flooding occurrences, despite having a history of terrible flooding (Little Stour and Nailbourne River Management Group, personal communication, March 21, 2015).

2.1.3 Purpose and Aims of Interviews and Focus Groups

The purpose of the interviews and focus groups was to understand participants' perceptions of the effectiveness of community resilience against the observed NRR security outcomes after a flood. The specific aims were to: (1) determine whether the identified list of community resilience elements is relevant to reducing the harm caused by flooding; and (2) identify any additional aspects of community resilience specific to British communities affected by flooding.

2.2 Methods

Interviews and focus groups were conducted from April to June 2015 in Kent and Surrey. Each interview and focus group lasted an hour. Interviews were one-on-one, while the focus groups consisted of two to nine flood-affected people. Interviews were conducted in the home of an individual being interviewed. Locations for each focus group varied, but a focus group was typically held in a village hall or a central meeting space conveniently near the participants. Each participant in the focus group or interview was asked to discuss various issues such as what happened to their community during the last flood, whether the recovery from the flood went well or not, and to name specific things that helped the community recover from the flood.

2.2.1 Inclusion Criteria

Participants were included in this research if they lived in Kent or Surrey, were affected by flooding, were 18 years old or over, and spoke English. Any participant who self-identified as being affected by flooding, which included the river, coastal, surface water, sewer, and groundwater flooding, was accepted in this study.

2.2.2 Participant Recruitment

Participants were initially identified through a snowball approach of passing out a recruitment notice within my supervisors' professional networks in the Kent and Surrey areas. Additionally, emails of this study's information sheet and consent form were sent to people who were affiliated with Parish Councils, County Councils, Borough Councils, the National Flood Forum, the Campaign to Protect Rural England, the EA, and Public Health England. Each email included a request to pass the information sheet and consent form to suitable and potential participants.

Following this approach, key stakeholders in the region, such as government officials, Parish Councilors, and Flood Wardens, were met. These initial meetings with the stakeholders gave a better background on the issues in the region and provided additional contacts and knowledge to recruit likely participants better. Additionally, interviews were conducted with local leaders in the parishes of Collier Street (Kent), Bridge (Kent), and Wraysbury (Surrey). Details were provided of the flood impact in the local areas and aided with spreading the recruitment notice around in the respective regions. The information that was given in the interview in the parish of Collier Street was vital in narrowing down the questions in the focus group discussions and learning the flood-related issues facing local homeowners, such as the local riparian owners defined by the EA as residents who own "land or property next to a river, stream or ditch" [16, p. 4].

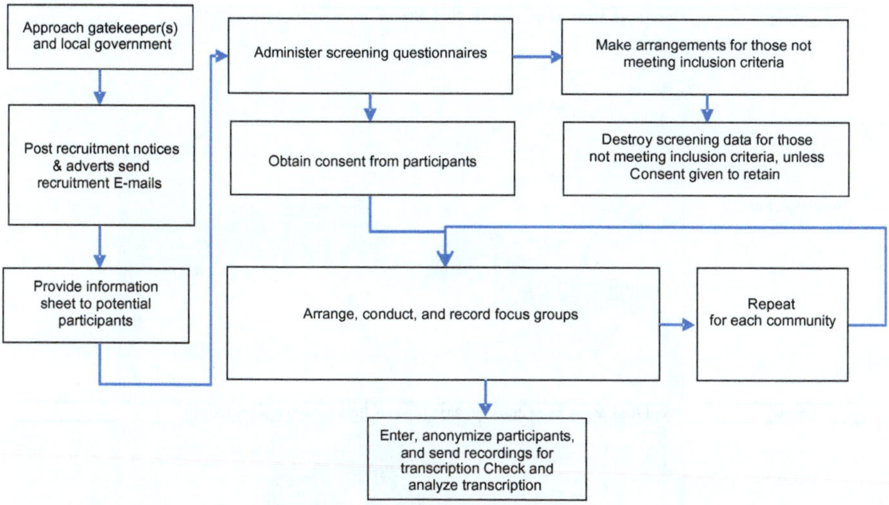

Fig. 2.1 Flow chart of recruitment

Initially focusing on recruitment in the Kent region, a press release was created to spread the word about this study and was emailed to several major media outlets in Kent. A gumtree advert was also created along with an online recruitment notice posted on the official website of the National Institute of Health Research's Health Protection Research Unit in Emergency Preparedness and Response at King's College London.

After assessing whether the interested participants fit the inclusion criteria upon initial contact, the study's information sheet and consent form were sent out. Each interested participant was also offered a chance to ask questions after reading through the information sheet and consent form. The approved participants were consulted on possible times and dates available. Each participant only joined one focus group. A flow chart of the participant recruitment process can be seen in Fig. 2.1. Figure 2.2 shows the EA's flood risk maps of recruitment areas [17].

2.2.3 *Number of Focus Group Participants*

Finding the likely point of saturation for the information given in the focus groups determined the number of participants included in this study. The practical and financial constraints of this study were also under consideration with this determination. The saturation point was determined when no new instances of new topics emerged during the collection period. This process was done by looking at items found in the focus groups using notes and lay summaries. An agreed number of groups and participants was determined during the collection period from these meetings.

Locality	**Flood Risk Maps as of June 29, 2015** [Source: Environment Agency (2015)]
Ton-bridge, Kent	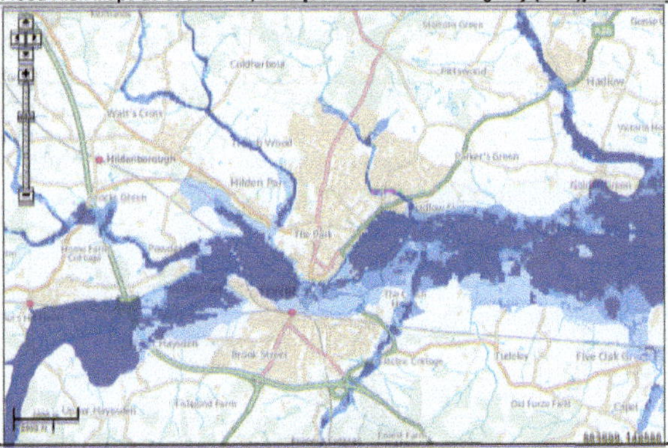
Locality	**Flood Risk Maps as of June 29, 2015** [Source: Environment Agency (2015)]
East-Peckham, Kent	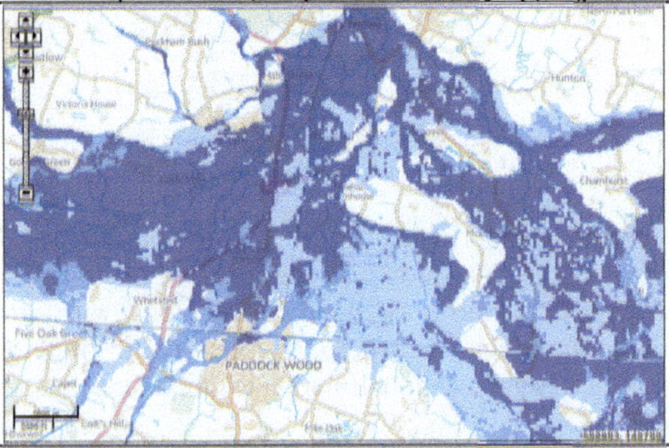
Locality	**Flood Risk Maps as of June 29, 2015** [Source: Environment Agency (2015)]
Bridge to Littlebourne, Kent	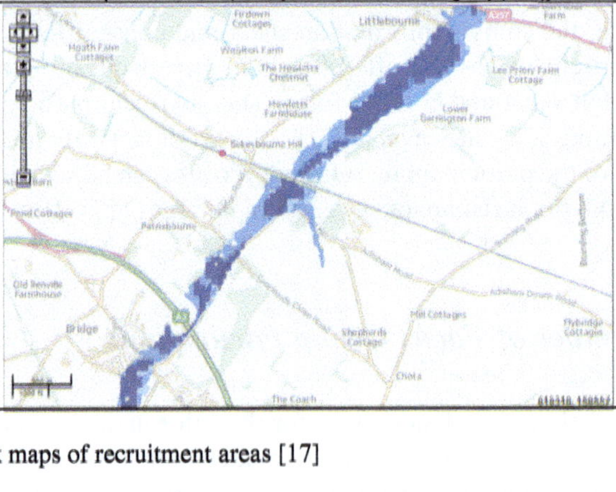

Fig. 2.2 EA flood risk maps of recruitment areas [17]

Locality	Flood Risk Maps as of June 29, 2015 [Source: Environment Agency (2015)]
Faversham, Kent	

Locality	Flood Risk Maps as of June 29, 2015 [Source: Environment Agency (2015)]
Chertsey, Surrey	

Locality	Flood Risk Maps as of June 29, 2015 [Source: Environment Agency (2015)]
North Staines, Surrey	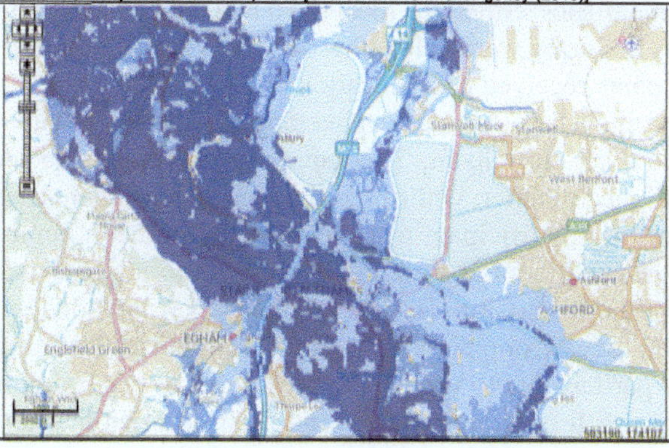

Fig. 2.2 (continued)

Locality	**Flood Risk Maps as of June 29, 2015** [Source: Environment Agency (2015)]
South Staines, Surrey	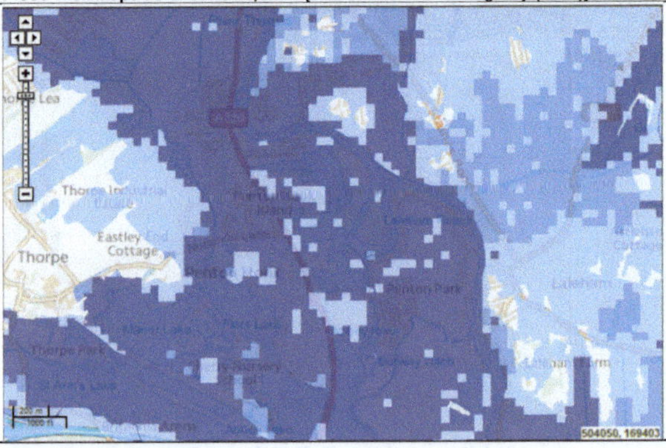
Legend:	**Dark Blue (high)** - High means that each year, this area has a chance of flooding of greater than 1 in 30 (3.3%). **Blue (medium)** - Medium means that each year, this area has a chance of flooding of between 1 in 100 (1%) and 1 in 30 (3.3%). **Sky Blue (low)** - Low means that each year, this area has a chance of flooding of between 1 in 1000 (0.1%) and 1 in 100 (1%).
	Light Blue (very low) - Very low means that each year, this area has a chance of flooding of less than 1 in 1000 (0.1%).

Fig. 2.2 (continued)

2.2.4 *Procedure*

Multiple groups were formed in Kent and Surrey regions. The groups in Kent were held in seven parishes, either at a local town hall or at a parish councilor's home. The focus groups in Surrey were held either at a town hall, community center, or a social club. Participants received compensation of £30 for their time and any expenses.

Upon entering a focus group, an envelope of paperwork with an assigned number was given to each participant. This envelope was folded to make a "name" card using the assigned number; while recording, each participant used this given number (e.g., participant 4, participant 6) to refer to each other's comments. Additionally, the moderator used the assigned number system to address and call on the participants during the discussion. Each envelope packet included an information sheet, consent form, focus group booklet, payment reimbursement form, and a blank sheet of paper for participants to write any notes. Prior to beginning the discussion, time was given for each participant to re-read the information sheet and ask any questions about the study or if any participant would like to leave and participate no longer. A signed consent form from each participant was collected before starting the discussion questions.

2.2.5 *Questions*

A detailed moderator guide was created and used in all groups to ensure consistency in the questions being asked and the order. This guide included a script, read verbatim for each group to introduce the project, the rules of the focus group discussion, and the confidentiality and withdrawal process. The moderator guide was developed and then shaped by the information given by the key stakeholders in Kent and Surrey and then edited.

2.2.6 *Format of the Discussion*

The discussion within the focus groups was split into three sections and corresponded with a focus group booklet given in the envelope of paperwork. The goal of using a focus group booklet in conjunction with the discussion was to help the participants gather their thoughts about the questions being asked and allow them to share their prior flooding experiences coherently. The first section asked about the number of years that the participants had lived in the community, the number of floods they had experienced, and an overall descriptive summary of their flooding experience. Participants were then asked to fill out the next page of the focus group booklet. The next section focused on the participants' own definitions of community resilience and examples of what they found helpful in their community against the flood impact based on their previous experience. Afterward, the participants were given a verbal explanation of the nine elements presented in Chap. 1 and Patel et al. [37]. Table 2.1 shows the verbal description given to each group. Participants were then asked to complete a table in the focus group booklet for up to five minutes. Participants were asked to provide their insights and examples based on their experience with each community resilience element. The last section asked participants to explain their responses provided in the table. If time remained, the participants were asked further questions in the guide, such as a question on the resilience of and aid to vulnerable people in their community.

2.2.7 *Transcription*

All interviews and groups were recorded, and these recordings were sent to an independent transcription company, Penguin Transcription. A written transcript of each recording was produced. As part of the first step of the analysis, the transcription was gone through line by line with the recording as a final checking measure.

Table 2.1 Verbal explanation of elements of community resilience given to each focus group

Element	Verbal explanation
Local knowledge	The concrete information, education, and experience acquired in relation to a disaster but also the specific learned information, such as basic life support techniques
Community networks and relationships	The connections in and out of the community and the nature of these connections
Communication	The risk and crisis information given and/or provided in relation to a disaster
Health	The wellness of the community and the health services serving the community
Governance and leadership	The public services and infrastructure by the government and community leadership related to a disaster
Resources	The basic supplies, such as food and water, and any tangible resources needed
Economic investment	The direct and indirect costs caused by a disaster and the revitalization of funds to eliminate the costs
Preparedness	The activities used to mitigate the effects of a disaster
Mental outlook	The views or feelings facing the uncertainty after a disaster

2.2.8 Analysis

A hybrid approach of thematic analysis was carried out on the transcriptions of each interview and group, using QSR International's NVivo 10 qualitative data analysis software for the coding process [6, 19, 20, 42, 44]. The hybrid approach used both inductive and deductive coding styles for theme development while using an overarching theme framework [6, 19, 20, 44]. The broad nine elements derived by Patel et al. [35] served as the overarching framework for the headings to initially categorize quotes. A new sub-heading was created under one of the nine headings each time if multiple quotes related to the same topic were found. There was information shared that was out of the scope of this study, such as ones that described the flood or focused on individual resilience rather than community resilience. These were captured as new headings but not included in the analysis. The information in the focus group booklets was used for background information and to supplement the quotes extracted from the transcriptions. The gender pronouns assigned to the quotes used in the result section were randomly determined. This assignment was to keep the participant's identity anonymous.

In the results section, theme names are in *italics* when appropriate. The number assigned to the interview or focus group is presented in brackets to illustrate where the respective quote originated.

2.2.9 Ethics

All data were treated in accordance with the UK Data Protection Act 1998. The interviews and focus groups were recorded and transcribed. However, none of the recordings could identify the participants as each participant was given and told to use an assigned number before the recording started. If names were used, the times in the notes were marked during the discussion and deleted in the transcription. King's College London Research Ethics Committee approved this research and all its documents used for data collection, ref: PNM/13/14-104.

2.3 Results

This study included a total of 77 participants. This total came from two interviews and 15 focus groups, which included two to nine people in each group. Participants included in this study were from the following parishes in Kent and Surrey: Bekesbourne, Bridge, Chertsey, East Peckham, Elham, Faversham, Hildenborough Littlebourne, Patrixbourne, Staines, Tonbridge, and Wraysbury. Six groups and one interview were carried out in Kent ($n_1 = 40$), and nine groups and one interview were carried out in Surrey ($n_2 = 37$). Table 2.2 provides the corresponding notation for each discussion to its respective locality.

Table 2.2 Corresponding localities to each discussion

Focus group (FG) or Interview (I) number	Locality
FG1	Kent
FG2	Kent
FG3	Kent
FG4	Kent
FG5	Surrey
FG6	Surrey
FG7	Surrey
FG8	Surrey
FG9	Surrey
FG10	Kent
FG11	Surrey
FG12	Surrey
FG13	Kent
FG14	Surrey
FG15	Surrey
I1	Kent
I2	Surrey

Table 2.3 Demographic characteristics of participants (N = 77)

Variable	Levels	Number of responses (%)
Sex	Female	48 (62)
	Male	29 (38)
Age	18–25	4 (5)
	26–33	4 (5)
	34–45	8 (10)
	46–54	13 (17)
	55+	48 (62)
Education	Left school without qualifications	14 (18)
	Secondary education (O-level/GCSE/A-level)	27 (35)
	Higher education (BSc, BA, or higher qualification)	36 (47)
Length in current community	<1 year	0 (0)
	1–2 years	5 (7)
	2.3 years	4 (5)
	3–4 years	3 (4)
	5+ years	65 (84)

Table 2.3 shows the demographic characteristics of the participants included in this study. Most participants were female (62%) and older than 55 years old (62%). The education background of the participants was mostly secondary education and higher (82%). Eighty-eight percent of the participants had lived in their current community for at least three years or more.

Table 2.4 displays the background information relating to the participants' past flood experiences. All participants had experienced flooding in their lifetime, with 73% of them only experiencing one or two floods. According to 96% of the participants, the winter floods from 2013 to 2014 were their recent flood experiences. Based on a five-point scale, most participants (83%) rated their last flood experience between mid to high severity.

2.3.1 *Meaning of Community Resilience as It Applies to Flooding*

Many participants listed characteristics related to the nine elements when asked about what it means for a community to be resilient to flooding. Some even used the terminology of "a process of adaptation or change," "bouncing back," or a combination of both to describe what they meant by the term community resilience. Still, most

Table 2.4 Characteristics of participants' past flood experience (N = 77)

Variable	Levels	Number of responses (%)
Number of previous floods	0	0 (0)
	1	31 (40)
	2	25 (33)
	3	12 (15)
	4	9 (12)
Date of the last flood	2001	1 (1.3)
	2002	1 (1.3)
	2012	1 (1.3)
	2013–2014	74 (96.1)
Self-rated severity of the last flood	1 (low severity)	4 (5.2)
	2	8 (10.4)
	3	17 (22.1)
	4	16 (20.8)
	5 (high severity)	31 (40.2)
	Did not answer	1 (1.3)

participants provided a broad but optimistic description of community resilience, such as "working together to make sure that you improve the situation or the future" [FG1]. No common or agreed definition was found within each group or across all groups.

Associating community resilience to *planning, protocols, and support*, participants said community resilience to flooding was:

> Sufficient measures in place, so that includes things like having the electricity supply that can withstand flooding; having support available; having a clear plan, which there wasn't having better communication about the flooding predictions over the next period of time. Potentially enough support as well around [FG12].
>
> Question of resilience in terms of helping each other, making sure that local residents, not Kent County Council and the Environment Agency or whatever, but those local residents know the people who may be subject to flooding and who may not be able to deal with it themselves [FG4].
>
> There should be a protocol for it [community resilience] if you live in an area known to be flooded everyone should be aware of the situation and what happens when it does happen [FG7].
>
> For me, the community resilience is firstly the willingness of the people in the immediate area to muck in and help one another, and I think having good lines of communication is absolutely vital [FG8].

Similar to *bouncing back*, participants also said:

> I think some of us have happily bounced back, but there are plenty that hasn't [FG9].

> We haven't got any choice. We've got to bounce back. You can't just put your head in your hands and go, 'Oh god, it's the end of the world.' You've just got to get back in with it [FG10].

Contrastingly, a different participant said,

> I don't think you bounce back; I think this has had an impact for the rest of our lives. I don't think you bounce back off it [FG10].

However, another participant described it more as *a process of adaptation or change* and said,

> [the] reality is people need to understand that recovery is not instant or overnight. It could be years. In fact, in the absolutely worst case, it could be never [FG6].

Combining both concepts, participants said:

> I think personally, the bouncing back is just wishful thinking. It's a much slower process, and I would say it's just a false way of projecting it because people get on with their lives as best as they can, but I wouldn't call it bouncing back quickly at all [FG9].
>
> I think you do bounce back, but you've changed inside because you're more conscious [FG13].

In a different group, a participant included post-flooding responsibilities and actions in his detailed response, saying,

> this time, there was a real feeling that, no, we can't just in a way bounce back, we might feel good, right, everything is fine, the sun's out, and it's the summer, but actually in the long term that didn't do us any favors last time. So those of us who'd experienced the previous flooding were very determined that we weren't going to let this go, and we went through about six months, I suppose, of a process with the local authority, the borough council, Thames Water, basically battling to get the proper investigations and some actual work done on the infrastructure, which is why we set up a Facebook page and we had a big meeting at a local community center, and then we had a steering committee [FG8].

2.3.2 Elements and Sub-Elements Found in the Interview and Focus Group Discussions

No evidence was found that new elements were needed to be added to the list generated from the literature's interpretations of community resilience. However, quotes from the discussions added evidence to the importance of each element and provided new details for sub-elements. Table 2.5 shows the key topics that emerged for each element of community resilience from the discussions.

2.3.3 Health

Several topics emerged when discussing the element of health. The participants' main concerns were with the *health services* provided to them and the *dangers of*

Table 2.5 Elements and key topics found from the interview and focus group discussions

Elements/key topics		
Health	*Resources*	*Preparedness*
• Quality and delivery of health services: physical health & mental health • Dangers of flooding, flood water, and sewage • Actions to ameliorate the sewage issues • Sewage and clean water concerns	• Wrong type or a lack of resources • Resources and essential goods bought by communities • Leftover resources and post-flood improvements • Resources that were said to help • Sandbags	• Lack of preparedness and plans • Grassroots efforts by the communities • Physical flood defences • Lessons identified and outcomes for action • Future needs to improve the community's preparedness and crisis management
Community networks and relationships	*Economic investment*	*Local knowledge*
• Community spirit • Isolation and abandonment • Community members buying or building their own resources • Community connectedness • Outside organizations helping communities • Community members helping each other	• Spending to prepare for a flood • Spending during the flood • Costs after the flood • Post-disaster spending • Government grants and post-disaster economic program • Insurance companies and the strain of insurance	• Taking actions using prior knowledge • Knowing who to call or ask • Community's history of rivers and floods • Community awareness of flood risk
Communication	*Governance and leadership*	*Mental outlook*
• Impact of media, filming crews, and tourists • Effective communication • Barriers to receive risk and crisis communication • Risk and crisis information from the local authority, officials, and outside organizations • Risk and crisis information from within or between communities • Mistrust of communication	• Having better leaders • Blame game type of politics and the lack of responsibilities by services and officials • Services and officials not being prepared or able to help communities • Community members giving crisis information to help with services • Tankers and workers • Lack of public involvement and support	• Adapting to a post-flood community • Positive and negative coping experiences • Previous flood experiences • Sudden shock of the flood • Trauma caused by the flood • Feelings of worried sick and not having any trust or faith • Outcome of long-lasting fear • Belief in never being able to recover and not having any hope for the future • Complacency for things to change after the flood

flooding to their health, such as the pools of *flood water and sewage* that remained after the storm. Anger at the public health response and the *actions to ameliorate the sewage issues* were other worries discussed by the participants.

The *health services* were stretched in some communities, which left many participants dealing with their emerging health issues on their own. One participant shared her first-hand perspective on the medical recovery in her neighborhood after the flood:

> Individuals who were physically injured due to the floods for whatever reason, whether it was due to the floods or simply the coincidence that someone had a pre-existing medical condition which was exaggerated by stress [of] the flood itself, went into a more serious condition [afterward] [FG6].

Most groups agreed that trauma caused by floods needs to be better understood and practiced in delivering quality care immediately after the flood [FG1–3, 5–10, 11, 12, 15, I1–2]. One participant shared his observation of his community after the flood saying, "actually, there are mental issues" in his community, especially "the effect on how I lived my life" [FG5]. This participant explained the impact on his life, such as the lack of access to transportation (i.e., his car) and the heightened risks of having to trek in the floodwater to retrieve needed supplies [FG5]. Many groups said they would be interested in learning more about the long-term effects of the flood on their well-being and health and adopting better preventative measures against the likely flood harm in their communities [FG1–3, 5–10, 11, 12, 15, I1–2].

When the topic of health was initially brought up, most participants immediately talked about the *potential dangers of flooding* to their health. One participant said, "health one is a big issue, even relatively we're fit and coming back in the evening after work, and it's dark, and the water is flowing, actually walking through that you could easily have an accident and slip over or whatever" [FG12].

The *dangers of sewage flooding* were also brought by participants when discussing the risks of flooding to their health. One participant said,

> if you moved too quickly, it [the flood] would sweep you off your feet, so it was quite dangerous to go out in it. And in a lot of areas, the sewers flooded, so you had sewage to contend with, and also you've got to remember that the things that live on the river bank, rats and stuff, there's a lot of disease in the water [FG8].

A different participant in the same group shared what he and most of his neighbors did in terms of combating the risk of disease and illness:

> In the end, we had disinfectant trays outside our houses because although the water wasn't in the houses, we had to wade through it in our wellingtons to get in and out. A number of people on our road had gastrointestinal problems, diarrhea, and sickness, and yet there wasn't information really widely being broadcast to say you need to take off your boots, disinfect your boots, make sure that you don't take any of this water inside – it might look clear but actually all flood water is contaminated, particularly in our road where there's sewage coming up [FG8].

Additionally, a few participants noticed children playing in the flooded water, which was mixed with sewage waste.

Any *actions to ameliorate the sewage issues* in the flood-affected communities were varied in speed and effectiveness. Based on several groups, the sudden onset of sewage problems, as one participant described the onset as sewage "starting to gush out of the manholes" on his street [FG3], was given a low priority. This conversation continued even when participants brought the sewage issues to the attention of the relevant utility companies. Substantial *sewage and clean water concerns* were said to affect and stress many participants [FG1–3, 10–12, 14–15, I1]. Participants described the difficulty of concurrently dealing with river and sewage flooding in conjunction with not having any facilities for clean sanitation [FG3, 10–12, 14–15, I1]. Many groups complained about how the utility companies and government officials did not understand these difficulties. Additional complaints came when discussing their experience with the utility companies, which many felt did not take an urgency with the issues nor know how to fix them. One participant said,

> Southern Water knocked the tops off their manhole covers, and they allowed raw sewage to go straight into that river. But we've been told if we contaminate that river on any works that we do, the EA will hammer us for it [FG10].

Another participant said, "[if] your sewage goes, you can't stay there [in your home], and nothing's been done about that, and that was a problem" [FG12].

2.3.4 Resources

Five key topics emerged from the discussion on the element of resources. The topics were *the resources that were the wrong type or not available*, *the resources that were bought during the flood*, *the resources that needed to be improved after the flood*, and *the resources that were said to have helped*. The aid of *sandbags* was debated in most groups for its purpose and usefulness against the flood. Suggesting other types of resources and how improving access to resources can help participants against likely flood harm were also discussed. For example, several groups agreed that better maintenance of the roads was required to help their communities after the flood [FG3, 9, 13–14]. One participant said, "once the flood waters had receded, the roads were in such a bad state" in his community which made traveling on them difficult [FG12]. During the flood, the access to the roads was limited, as the streets were often flooded,

"you couldn't get in, you couldn't get out," but cars kept trying to "get through the water that was far too deep for them, [and] breaking down" [FG13].

Participants felt forgotten because their community was provided the *wrong type or lacked resources*. Many participants blamed not having the appropriate resources in the community because their community was not being vocal in demanding resources like others that they have seen in the media [FG2–3, 6–9, 10–14]. In another group, the lack of sandbags made the participants feel isolated [FG13]. A few participants in a different group said they did not get any resources to prepare against the flood and were not even "aware of sandbags" or "how to get sandbags" in their community [FG8]. Some discussed their lost access to their showers, washing machines, and bathrooms, and many of them were told that emergency portaloos would be delivered near to them by regional leaders. However, these emergency portaloos were challenging to use because the participants were so flooded and could not easily reach them [FG14].

Many participants and communities collectively *bought needed resources and essential goods* during and after the flood. Neighbors in the communities were giving aid to each other, such as providing lifts to the local shops and even to and from the train stations [FG7–9, 14–15, I1–2]. Still, some participants questioned why more government resources were not allocated to prevent future flood impacts in their communities [FG3–5, 7–10, 14–15, I1–2]. One participant said the government should "spend the money on [flood] barriers for us instead of these useless people" managing the recovery efforts [FG10]. Another participant suggested that there should be some sort of an exchange or store where "you can go to get this kind of stuff [supplies for flooding], and even if you don't buy it [the supplies], you borrow it and then take it back" [FG12]. This exchange idea had many participants in the same focus group thinking about maintaining and storing the recently bought resources until these resources are needed again for a future flood.

As time went on since the onset of the flood, participants said they were becoming more concerned about *the resources they needed to improve against a future flood and what to do with leftover resources* that they bought or received during the last flood. A big concern for some participants was solving the blockage issues on the drainage and pipes in their homes and communities [FG1, 3, I1–2]. One participant said, "one good thing came out of it [the flood], we all had new drainpipes put in underground" the community [FG14]. Another concern was being better informed about where people should go to receive preparedness and emergency resources if another flood occurs. One participant said he and his fellow community members should

> know where we could go if the government or the council had this stuff, and if something happens, you go there, you can go and get it [FG6].

Another participant was able to find and use the local resources available to her but did not know where to return them after the flood. She said,

> after the event, we didn't know what to do with them, so you were just left there with these things that, again, you couldn't carry. Some people have still got them there; other people just smashed them open, so you end up with all this money spent on the sand to fill them that's just wasted, whereas there should've been a place where they go to, and they're kept for the next time this happens [FG12].

Many participants were in conversations with their local councils concerning which resources were needed if there was another flood. One participant had:

> been to several big meetings at the council and stuff where they've said, 'What equipment do you need, what can we get for you to be ready for next time?' and things like that, and we've given them lists and things, and we haven't seen any of it [FG6].

Participants said there were many *resources to help them* during and after the flood. Participants in one group echoed the feeling of being "lucky because of power and heat and that stayed on," which allowed them to remain functional in homes in the middle of winter [FG12]. Another group of participants was thankful for the generous food parcels from local businesses and charities in their locality [FG14]. Sandbags were also said to help participants, as one participant said she was "well-supplied with those [and] well supported with that" [FG5]. Still, not everybody was well-supplied or agreed on the sandbag's usefulness against the flood [FG1–15, I1–2]. *Sandbags* were the most discussed resource item across all participants [FG1–15, I1–2]. Many participants debated in the groups about the value, especially if the resource becomes available to them after the initial flood of water hits the community. One participant shared her opinion that "a badly assembled sandbag barrier is no use" and "it's certainly no use if your road's already flooded" [FG6]. She shared that when sandbags were made available to her, it was too late to help her. In the same group, another participant compared the resource of sandbags to a "sticking plaster for a situation" and claimed the sandbags were not the answer to their long-term issues in the community about flooding [FG6]. In a different group, one participant expressed the high amount of urgency that his community had "to get some sandbags sorted out" because he was "rushing around trying to find somewhere where we could get sandbags" to his community before the flood hits [FG4]. From their debate-like discussion, one group finally concluded that having sandbags was more for making the residents in the community feel good about their preparation against the flood rather than if the sandbags would prevent any severe flooding [FG6].

2.3.5 Preparedness

For the element of preparedness, several vital topics emerged from the focus groups. One was the community's *lack of preparedness and plans* to face a potential flood. Another was the *grassroots efforts by the communities* to prepare themselves for the next flood. Participants in every group organically brought up a discussion around *physical flood defenses* in their locality. Several lessons were identified, and *outcomes for action* were shared in the groups and what would *help the participant's (respective) community for the next flood.*

Participants echoed a consensus when discussing their *community's preparedness for a flood.* Most participants felt more reactive to the flood than anything else. Another participant described the preparedness in his community as a "lack of attention," stating "it's like out of sight, out of mind" [FG8]. He went further and said that people "only do anything about it [flooding] or think about it when it happens again unless the people go on and on and on about it" [FG8]. The groups discussed the benefit of being aware of the need to be prepared for a flood in the community. One participant said that "if a community is aware that it's at flood risk, it's more mentally prepared and hopefully more physically prepared as well" [FG2]. Another participant compared this risk awareness between people living near and far from the river. She said, "if you were by the river, you already have a notion that something may happen, but if you're away from the river, there wasn't" [FG9].

Specifically, *the lack of plans in the community* to anticipate flooding was an emerging topic. Participants were astonished by their community's lack of plans for the previous flood and for anticipating a future flood [FG1, 6, 8]. One participant said she and her neighbors "all realized that it's not a once-in-a-lifetime event" after learning about their community's lack of preparedness and plans for flooding [FG1]. Another participant highlighted the fact that the "last published full strategic flooding plan was published in 2009 [and] there was a minor update in 2012," which shocked the participants in this group [FG6]. "Lack of practical preparedness" was echoed in several groups, with one participant saying that "it's the unglamorous bits of the work that need to be done" in his community [FG8].

Grassroots efforts by the communities were discussed as a better option to prepare for a future flood. Many participants were aware of selecting flood wardens and creating a community-based emergency committee, but their effectiveness was questioned. One participant said,

> I mean, I don't think we can make plans for that; just as I was saying, having an emergency committee is fairly useless in an emergency situation; the good thing about it is being aware of what we should be able to do if things happen, I think that's all you can do really [FG1].

Another participant said, "we're supposed to have flood wardens in the village; again, that didn't seem to materialize around where we were" [FG12]. However, some participants were hopeful that these efforts would be helpful in the future. "A little bit of protection and people that are supposed to be there and are trained to have knowledge of what needs to be done" was said about having flood wardens present in one participant's community [FG13]. Another participant highlighted the benefits of a community-based emergency committee as "you know what people around you are capable of, how they can help next time around, [and] how we can all prepare ourselves" [FG9]. Furthermore, many participants agreed that the self-help created by the neighbors during the flood could be sustained afterward through grassroots preparedness. For example, one community created "spreadsheets of everyone's telephone numbers and email addresses, and in some cases, their medication" along with indicating "who couldn't look after themselves" [FG9]. Because of this, their community was:

> just so much more prepared now, and we know how to go about it, and we know that we can't expect to rely on the council doing anything or perhaps even emergency services until it's serious [FG9].

Physical defenses for the community and households were discussed in every group. The discussion mainly focused on the effectiveness and the decisions of individuals buying their own protection. In one community, the "alleviation channels, flood defense, was actually working beyond its maximum capacity that it was built for" [FG1]. In another community, the flood was reported to be primarily caused because "senior personnel could not be reached for advice" on whether water should be released into the alleviation channel" [FG12]. The staff member on duty, according to the participant, decided to run water into the alleviation channel, which was already overwhelmed due to poor maintenance and infrastructure [FG12]. Other communities also reported poor maintenance and infrastructure for their physical defenses, such as with the waterways and assigned flooding ditches along the roads and river streams. Lack of maintenance and infrastructure was why more participants decided to buy or organize their own defenses. "After the last flooding event, people have started to take measures to make sure that their houses are more resilient," said one participant [FG5]. Overall, the groups shared a general feeling of resentment about physical defenses, especially for those organized by the government. One participant said,

> there's always lots of promises, lots of things they're going to do, but once it's over, it's forgotten about it until it happens again and so nothing to me seems to get done in a proper fashion. They dig a few ditches and think, oh, that'll do. But it's not enough. And not enough is done not just for this community, for many a community that suffer all the time, every year, more than what we do. And still, nothing is done, and to me, it's a big thing [FG13].

Essential lessons from their experience with flooding were identified and suggested several *outcomes for action*. In all groups, the initial learned action was to be less reliant on government and other organizations outside of their community. Many groups organically empowered themselves to remember that they have been through and survived past floods without such support and could endure future floods on their own. Participants said:

> Once you've lived through it, you just know what you've got to do to get ahead of the game a bit [FG2].
>
> One thing we learnt from the first flood 'cause we just had lots of stuff there, and it just got ruined, so we then bought plastic things, and we move stuff up as soon as you know [FG12].
>
> People have actually learnt from the last experience, and they've put in walls to stop water coming through from the neighboring fields and things, so people are starting to make themselves more flood resilient by taking those actions to flood-proof their homes [FG5].

Even sometimes, putting up flood defenses may not be effective; as one participant said,

> it hadn't helped at all because it [the water] had gone so high it had gone over these things that were on the door [FG7].

Integrating the successes and fixing the failures from previous floods were crucial to the participants. They stressed that their community and government agencies should practice this openly [FG11]. One participant said,

> they just think it'll go away and that next time around, everyone will have forgotten what happened the last time, and that's not how it happens; the whole point is you're supposed to learn some lessons from this to prevent it the next time and any mistakes that were made identify them, not blaming people [FG12].

Another outcome for action was increasing the training and transferring the knowledge from previous floods in the community and across relevant organizations. Participants said:

> The important thing is the body, is the parochial charities, the parish council, or similar, or borough, where that knowledge is kept as much up to date, that that training is kept most up to date so that they know what they need to know in the event of a flood [FG10].
>
> I don't like to say we're on our own, but I do think we are the first and last defense, is our own community. I think educating the younger generations and training them and preparing them is critical, and not just for flooding but for any disaster [FG11].
>
> There is plenty of time to react, and what we need to know is that our civil servants understand what they need to do when we're in trouble [FG5].

According to the focus groups, these lessons will help the participants and their community better prepare for a future flood.

When identifying lessons from the flood experience, the topic of *future needs to improve the community's preparedness* emerged from the discussion. Ideas on how to improve future preparedness and crisis management were shared. A few participants shared a vision of having better transparency in flood preparedness plans [FG2, 5, 7, I1–2]. For example, a suggestion was to have a welcome pack for each household in a community and include relevant information to prepare for a future flood:

> if, for example, you're going to get flooded, these are all the things you need to be aware of, you need to go and make sure you've got insurance, these are the services available to you [FG7].

Having lists of resources and contact information could also be included in these welcome packs, which could also be available electronically to ensure the knowledge is easily accessible and frequently updated. Another suggestion was having a call-in phone service once the flood risk warning has been issued. As one participant suggested, this service could be "like the Samaritans" [FG5]. She continued to describe the service as you can "have somebody that you can ring in the middle of the night and say, 'I'm really frightened, the flood water is rising here'" [FG5]. This service could help reduce the heightened anxiety of experiencing flooding and feeling all alone. A few groups suggested having official debriefing events in the community [FG3, 5, 7, 10, 11, I1]. These events might effectively identify global lessons from the flood experience and begin the healing process for life after the flood in the community [FG3, 5, 7, 10, 11, I1]. One participant summarized the value of identifying outcomes for actions from past flood experiences as to how change can come:

> Its value [of memories and experiences] should be talked about and shared with others because this is how change can bring about [FG5].

2.3.6 Community Networks and Relationships

Most frequently discussed in all groups was the element of community networks and relationships. Participants shared their experience in this element by giving examples of *community spirit, isolation and abandonment, community members buying or building their own resources, community connectedness, outside organizations helping communities*, and *community members helping each other*.

Overall, many participants described similar thoughts about the importance of community networks and relationships. Many participants felt that the community networks and relationships were "the greatest asset throughout the flooding and post-flood" [FG9]. Many participants also described their community as cohesive with a *high community spirit* that helped them deal with the flood. This closeness carried on long after the flood and was seen as much-needed support for life after the flood. Participants said:

> There was a togetherness because we knew no help was coming [FG5].
>
> The community feeling was amazing, and it [post-flooding support] was all community based [FG13].
>
> Community networks and relationships probably is the main one for me because if we didn't have what we had, then I think it would've been really hard to get through [FG13].
>
> Beforehand, we knew what people's names were and knew people well enough to nod and say good morning, but that was it; whereas now we know more about people, and we chat, we help each other out, and yeah, people are much more cohesive really in the street [FG12].
>
> When everybody comes together, it makes it easier because everybody's experiencing the same difficulties and fears and worries [FG13].
>
> It makes you feel like you've got a little shield around you because you're all in there together [FG13].
>
> We knew they came together at a time when they were needed; we remain friends with so many since then [FG11].
>
> The community spirit got everyone together [FG9].

Describing how important the community response to the flood, participants said:

> I would say without the community response, then we would have gone under even worse, the whole village, and it was really the community themselves that took charge until all the emergency services and everyone else jumped on the bandwagon and came and helped out [FG11].
>
> It was the community that did all the help [FG11].
>
> Feeling of there were people around that were concerned about what was happening, and in our immediate vicinity of neighbors, we all checked on everybody else [FG4].

However, there were a few participants who could not help or did not like this newly developed closeness and said:

> You can't do anything when you're surrounded by water, and it's running fast through your garden and taking your car away, whatever. But you know that your neighbor is in exactly the same boat [FG9].
>
> It shouldn't take a disaster like that to get a community feeling together because I don't feel the community feeling's been the same since the flooding [FG13].

Stemming from community spirit was *community members buying or building their own resources* and *community members helping each other*. A participant said his flooding experience was a "very encouraging experience because it shows that people tend to rush and help if and when needed" [FG1]. A clear example of this was given by a participant who said, "people were outside the church just passing by, would be making up sandbags for people to collect, so that's an example of really good community spirit" [FG7]. Another participant said, "half a dozen of us actually filled our own sandbags and distributed those [to others in the community]" [FG1]. In the same group, a different participant described a different exchange:

> Someone came in with a tractor, and someone I hardly knew to say hello to came with a tractor to move the sandbags, and the farmer supplied the barn for us to keep the sandbags in and to fill the sandbags [FG1].

Besides sandbags, other resources, such as water pumps, portable flood gates or barriers, and portable toilets, were also bought or built by communities [FG3, 5, 8]. Another example was dragging neighbors' trash and recycle bins out of the flood zones for trucks to collect them [FG10].

On the topic of community members helping each other, many groups agreed with one participant's assessment of the flood experience: "people pulling together, I think, is the main thing" [FG13]. Examples of "people pulling together" were said to be:

> People worried about the electricity going off and things like that, so people got together and tried to find out what was going on [FG13].
>
> People who don't even know each other, people that haven't been living on here for a long time were all chipping in [FG13].
>
> People were asking what people needed and then trekking through the floods to go and get shopping and supplies [FG13].
>
> What was good in a way was the sense of community, everyone, kind of once you were in that situation, did pull together, people with boats helping people and people generally working together and realized it was a good thing to do [FG12].
>
> We pulled together and were happy to be phoned up and take any messages or help or whatever or bring food round for people if they couldn't manage to get it [FG12].

Additionally, having a local center or meeting point was helpful for community members to give and receive help. One participant said,

> having the community center was a big, big help because people knew they could come here. There were teas and coffees and sandwiches being made, and it's a place to gather to find out exactly what was going on with everybody else, [and] to see if anybody needed any help they'd come in here because that was the focal point [FG13].

Community members even went door-to-door to check on others:

> In our road, we had three residents who were going up and down every day knocking on people's doors making sure they were alright [FG12].
>
> I had my chest waders, so I was going down and pushing the cars out of the water [FG13].
>
> Our neighbors all came in with their brooms, and we got a pump, and they were sweeping the water into the pit, and it was being pumped out as fast as it was coming in [FG2].

All groups noted that many of the positive outcomes that happened in the community were because of the efforts of the community members [FG1–15]. Many participants discussed and complimented specific individuals in their community who made a difference to them:

> Our neighbors worked well, as a community worked very well together [FG7].
>
> We would've not survived without certain people within this community, to be fair. There were certain individuals within this community that made the whole thing bearable and cope-able with. If it weren't for those certain people, we would've struggled really badly [FG9].
>
> What worked well was the community themselves, how the community quickly came together to help those that were in serious need of help, like the elderly or disabled or young mothers with children who really needed a hand to get out of their properties and stuff [FG6].

Another key topic was *more community members being connected* to each other due to the flooding experience. Being connected to neighbors who helped and now knowing who to contact if a flood happens in their community were two points that many participants gave when discussing the connectedness of the community [FG1, 3, 5, 6, 9]. Participants said:

> We've got to know so many people since the floods that we didn't know before, and yet they're sort of neighbors [FG11].
>
> After everybody's houses were fixed and we moved back in, we had a big party in the street to welcome everybody back home, so that became quite a community thing really in our road, and it's something that we're doing every year now, so people know each other. So that was something quite positive that came out of it [FG12].
>
> The neighbors were the best thing, and we made good friendships, and we supported one another, and it was the best thing [FG9].
>
> I think if we didn't have floods, I would still only know just my immediate neighbors and maybe a couple of more people on the road. So, the positive things of the floods for me, the most positive one is that I got to know people in the community and through their help and my help with them [FG9].

Instead of feeling like a connected and cohesive community unit, some participants felt that they were forced into *being isolated and abandoned throughout the whole flooding experience* in the community. Participants said:

> Two-thirds of the people in the road had moved out by that time [time of the floods] because it affected them, some people just couldn't get to their house, other people's houses were flooding [FG12].
>
> If we had not had the wherewithal to help ourselves, we would've been totally flooded [FG10].
>
> It is very, very depressing when you're on your own [FG15].

Another reason was the divide created by the flood between the community members who were flooded and those who were not. According to a participant, the following best explained this divide:

> Because if you haven't been flooded, you don't care about it particularly, you're not bothered about it, but actually there's plenty they can do. They can help deliver sandbags, they can go and help the elderly people put their stuff up, but they just don't think about it, they're not aware of the problems that other people are going through when a flood happens [FG2].

Because of this and not wanting to cause trouble, confrontation, or complaints with others in the community, some participants felt the need to keep to themselves during the floods and afterward [FG10]. Additionally, newcomers in the neighborhood reported feeling disadvantaged, as they thought they were on their own:

> Basically, we're newcomers. We haven't built up strong relationships. I mean, we've made friends and neighbors and whatever, but I think everyone was just relieved that it wasn't them, to be honest. So, there was no support for each other...It's an everyday thing with the neighbors; I mean, every day when we walk past them, and it's uncomfortable constantly. It's not going to get any better [FG10].

Another focus group also saw this feeling of isolation and abandonment [FG11]. Three participants discussed the level of community engagement and support when adjusting to life during and after the flood. They said the following:

> I don't think we did well in the post-flood caring for one another. We were far more open and able to talk during the floods, and then once everybody goes home, you start to have to feel you have to look after yourself, and therefore you don't want to ask somebody for help. You go back to feeling, 'I have to deal with this by myself,' and there is nowhere to go [FG11].
>
> We used to meet every day of the floods, we had dinner together, and all of a sudden [no more of that]; so it's post-traumatic, you feel alone. You suddenly think maybe people don't want to talk about it, they don't want to know, and I just think, to be perfectly honest, I don't think we know really, the emotional damage that it's done [FG11].
>
> I think that we could have done a lot more. We could have got together after the flood, which we didn't do. As a community, we could have got together and helped some vulnerable people really [recover from the flood]. And I think we were all so pleased that the water went [away], that we didn't think about anybody else, we were very selfish in our own homes [FG11].

Despite some communities feeling abandoned, some communities *received help from outside organizations*. For example, four groups discussed the importance of volunteers, several big charities, such as a faith-based non-profit organization, and local supermarket stores, such as Tesco, providing and delivering food to their community center [FG6, 7, 9, 11]. In one of these groups, one participant said,

> there were a couple of others [outside organizations], and they literally just came in, they bought us] lorries and whatever we wanted. They still to this day keep in contact [and ask], 'do you want anything, do you need anything?' We help them now with other fundraising things, but without groups like that, groups like the Lions Club the Rotary Club and that giving us financial aid, so we could go out and buy waders, health, and safety equipment that we couldn't have done the job [of helping our community against the flood] without [FG6].

Another participant in these groups discussed how volunteers helped in the clean-up efforts in the community, such as "cleaning up graveyards and gardens" [FG7]. Another example given by a participant was that his community received guidance on what to do during and after the floods from outside organizations [FG12].

2.3.7 Economic Investment

Discussion on economic investments and how they played a factor for the participants was separated into three time points: before, during, and after the flood. Before and during the flood, the focus was on *spending money to prevent further damages* caused by the flood. After the flood, issues of *post-disaster costs and spending, government grants and post-disaster economic development*, and *insurance companies* were discussed at length by most groups. Almost all groups acknowledged the importance of financial investments and emphasized how spending money after the floods helped the flood-affected residents and communities [FG1–5, 10, 12, 14, I1–2]. For example, when discussing the damages caused by the flood, one participant said, "we need money spent here to sort this out" [FG10].

Participants said there needs to be *more spending to prepare for a flood* [FG3, 5, 10, I1]. The discussion focused on the costs of removing all the trash collected on the riverbanks that impeded the river's flow, along with the potential of dredging the river.

During the flood, many participants discussed the out-of-pocket expenses they had to sustain to keep functioning in their homes and communities. As one participant said, "there was no immediate money available, all those pumps [used during the flooding event] we had to pay out for by ourselves" [FG10]. One participant said his family was "thousands out of pocket, thousands" [FG10]. Another participant talked about the loss of income and lack of compensation to the local farmers; "over 2,000 acres flooded here, and there's no compensation for that at all, for farmers" [FG1].

After the flood, many participants tried to develop better flood prevention in their community or even leave the community, but the economic costs slowed down their progress. In one group, a discussion started about raising a footbridge that caused severe downstream floods in 2013 and 2014. One participant, who was initially skeptical about the effect of the footbridge, was worried about the costs on the community members:

> Raising the footbridge is fine. I can understand that they're going to raise the footbridge. But we're going to pay for that as a community. It's no funding from the Government or the EA [FG10].

Another participant talked about how people have unsuccessfully tried to sell their homes after the flooding event: "one person in the road has been trying to sell it for a year now without any success, so there's the worry about that in the future" [FG12]. Another participant noticed "more houses up for sale" in her community, but "the buyers have pulled out because of the history of flooding, so it is going to affect the properties quite badly" in her community [FG1]. Another participant talked further about the housing market:

> I know my property is on the market at the moment, I know I'm going to get questions about flooding, and I'm going to have to reply to them honestly. When I bought this property in 1998, it wasn't even on the form, you didn't ask about flooding, so I think there is a whole mindset change [FG5].

According to the groups, some post-disaster spending, especially from the government, did not help reduce future flood risks. Many participants saw their government spending on remedial work in their community that would not help reduce their future flooding risk situation [FG1–3, 10, 12, 14, I1]. Post-disaster economic development in the community was very briefly discussed in the groups. It was explicitly mentioned to a participant's housing situation [F4–5], a community's infrastructure and businesses [FG2, 8, I1–2], or insurance [FG5, 10, I1].

On the other hand, *government grants* were discussed in detail. Many grants were designed to help participants and communities build better barriers to prevent future flooding as a *post-disaster economic program.* One positive was that many participants could use the funds to empower their flood wardens with vital necessities by pooling their awarded grant money together [FG1–3, 10, 12, 14, I1–2]. For example, one participant said they had

> given a grant for the flood wardens for waders, chest waders, each warden will have its own road to control in future flooding, which we hope we don't have [FG13].

Another participant suggested that government grants should have been combined so a specialized consultant could visit and review the community's measures for flood prevention because

> the money that we were given off the Government, the £5,000, basically is just enough for us to buy materials. We couldn't afford to have a consultant come and say what we needed. We haven't got the money [FG10].

Similarly, another participant said, "government needs to stop giving silly grants of £5,000 to all of us and actually use that money and do a decent flood defense system" [FG12]. Another participant concurred and said,

> it's a complete waste of Government money and tax money, frankly, and they [the government] could've used that to do a decent flood defense system [FG12].

Another issue brought up by participants regarding the government grants was the long process of applying, receiving, or hearing back about financial aid. One participant talked about the length of fitting flood defenses in their home and the process of receiving financial assistance from the government grant:

> We've been promised [by EA] that they [EA] will come around and fit defenses. We applied in January of last year, and I've still not had anyone knock on the door except for a survey which we paid for ourselves; we got the money back [for the survey], but we're still waiting [FG11].

Many participants did not have free time to apply for it because of work and/or dealing with life after the flood. One participant said she tried earlier in the year but then was told to forget it because of the process. She said, "I tried asking for it in February/March of last year and was told forget it, we're all too busy, and it's just been done now" [FG2].

Some participants were unaware of their eligibility to receive the financial grants or any announcements of such assistance. Many participants exchanged information with other participants to acquire new details on the grants' deadlines and their help in applying for the grants [FG2–4, 7–9, 14–15, I1]. When asked whether the information about the grants was easily accessible for everybody, one participant said, "only cause of the people we knew at that time," and another participant in the same group said the local "MP put me onto it" [FG3]. Even when made aware of the government grants, many participants said they did not know how to apply for them. One participant said,

> I'm 26, and the forms lost me. They were asking me things about a property that was built 40 years before I was born – how am I supposed to know that information? [FG7].

Another participant explained further about the grant rules by saying, "at the end of the day, the government rules were interpreted differently by the different councils, so Runnymede Council said that unless it was actually something to do with your house, then it wasn't eligible" [FG7].

Lastly, *insurance companies and the strain of flood insurance* were discussed in all groups. One participant said that "it takes a lot of hassle dealing with insurances, dealing with everyday life" [FG9]. Another participant in the same group elaborated and said, "suddenly you feel that there are a lot of people and businesses or organizations who are making money out of your misfortune, instead of everybody running around to help you truly, they are just making money out of your misfortune" [FG9].

2.3.8 Local Knowledge

Overall, many participants perceived the element of local knowledge as an underutilized element based on their experience. This perception was best seen with emergency services and government officials who "did not always listen to us," as one participant said [FG12]. Examples of *taking actions using prior knowledge*, *knowing who to call or ask*, *the community's history of rivers and floods*, and *community awareness of flood risk* showed the benefits of having strong local knowledge to increase their community's preparedness and response to flooding.

Many participants felt they had a better grasp of what to expect and do with the floods by *knowing who to call or ask*. Having a list of phone numbers that people can call for information and support was "extremely helpful" [FG5]. For example, a participant said, "I know the phone numbers and who to go to," and another participant said,

> I certainly have phone numbers now that are on a fridge magnet tucked around the corner because in the back of my mind, as soon as the summers over, you're thinking [of flooding] [FG10].

In a different group, one participant said, "I remember you [another participant in the group who was a neighbor] sitting there thinking 'thank god we've got a phone number for A, B, and C'" [FG2]. Another participant said, "I did feel that if a problem happened [like flooding], that there would be people that I could contact" [FG3].

Besides having a phone list, other responses included going up to a local center or church to ask or asking the local councilor [FG3]. A few participants said they did not know who to call or ask during their experience and, as a result, found the floods difficult to deal with [FG7]. Looking back, a participant said she would have done things differently if she knew who to call or ask:

> So, to have some information to know how to handle something would've been brilliant, just to know who to call, what should I be doing? Because I'm sure, there's lots of things I would have done very differently had I known beforehand [FG7].

Using the example of a "community welcome packet," as mentioned in the element of preparedness, participants suggested that communities should

> at least have some information in a packet saying, 'welcome to our community, here's a contact list, here's some things from the past that people have faced' [FG7].

Many groups, in theory, thought *taking actions using prior knowledge* were needed in their community and discussed how local neighbors and officials could better implement it. This conversation signaled overlapping ideas with the elements of preparedness, communication and governance, and leadership. For communities to take action to relieve flooding in their community, the community's prior knowledge needs to be shared and heard by local authorities [FG1, 5, 11. I1]. For example, one participant reported that the local authorities in his community did not act on nor listen to his community's prior knowledge about flooding in his area [FG11]. After reviewing the plan implemented in his area to relieve the flooding, he strongly believed that the plan by the local authorities would not solve anything in his area [FG11]. All participants did not see this lack of confidence. Another participant said, "I felt confident that it would be fine because of the new works [against flooding] that had been done" [FG1]. These new works were based on recommendations by this participant's flood warden, who shared the community's collective knowledge of past floods [FG1].

The *community's history of rivers and floods* was passionately discussed in most groups. From frustration about not being heard to having pride about a rich communal history, participants shared similar examples in the element of preparedness. The distinction between the elements was that preparedness included lessons identified from floods in recent years. However, examples seen in the community's history of rivers and floods were deeply rooted in the community for generations and drew on historical information going back much further. For example, a participant in Kent said knowing the flooding in 2001 helped him and his community for the floods in 2013 and 2014 [FG3]. Another participant in the same group agreed and said, "my local knowledge of our area is that we know the stream" [FG3]. Furthering this point, one participant said,

> we spoke to a lady, and she said for years, she's lived in this area for years, and she said every year they [local authority] would go down and clear all the greenery, any branches that have fallen in and they used to dredge it [the river] [FG10].

Another participant discussed the historical importance of the frequency of the floods in the community, saying it is "really important point to capture" [FG5]. Her point was that there was a "big gap [between floods] from 1960 to 2001 and then the much shorter gap to 2014 [floods]" [FG5]. Because of this, she said that most of her community, especially those who had been in the community for a long time, remembered the past flooding and knew they would be dealing with it, especially the riparian owners, who have property near a river, stream, or ditch [FG5]. Echoing this was another participant who said,

> if you live on the river, you're prepared at some point to be flooded. They're used to it, they know how to recover from it, they have groups along the river, residential groups, and they have their own systems set up [FG6].

Because of some community's rich history of constant flooding, few groups discussed an individual or a group of people who were vital due to their expert knowledge [FG1, 3, 5, I1]. Talking about one of the local leaders within the community who led the community's flood crisis team, a participant said,

> he's worked and lived in this village all his life. His father was the flood warden before him. He's a builder. Therefore, he's probably built 90% of the extensions in this village. He's put everybody onto Mains drainage. He knows people. He knows if they're ill [FG11].

An example of a local and collaborative river management group that has one of their goals to retain "a history [of flooding] now that goes back over the last 15 years" was given as well [FG3, 5, 10, I1].

Community awareness of flood risk ranged from participants saying everyone knew to hardly anyone knowing about their flood risk. For those that knew, many either prepared with the best intentions or left the area. Armed with the knowledge of daily information on rainfall, one participant said, "we knew on a day-to-day basis what was expected, which was excellent, even if we didn't always realize what that meant" [FG1]. Because they knew the risk, one participant said, "that's why my neighbors just disappeared; she actually went to Holland" [FG10]. For those that hardly knew about the risks, one participant said there

> should be a protocol if you live in an area known to be flooded everyone should be aware of the situation and what happens when it does happen [FG7].

Another participant said there needed to be a more significant push to make "people realize that they have to take precautions to make their house more flood-resilient themselves" [FG5].

2.3.9 Communication

All groups mentioned the communication before, during, and after the floods. The discussion emphasized the essential risk and crisis communication. It included the same categories for risk and crisis communication: *barriers to receiving communication, information from local authority, officials, and outside organizations, information from within or between communities*, and *mistrust of communication.* Additionally, several groups discussed the *impact of media, filming crews, and tourists* on the flood-affected people and areas in their community.

All participants in one group agreed that the "biggest problem is communication" [FG11], but it was also echoed by many other people when asked about their flood experience [FG2–4, 7, 10, 15, I1]. Most participants felt a breakdown in receiving information on what to do or how to act and blamed it on officials not communicating effectively with them. Participants mentioned *effective communication* as a vital thing needed and that it consisted of "good and clear" messaging [FG1]. Only one group explicitly discussed effective communication, while the others did not clearly define or even spontaneously mention it. However, a few participants described good communication as speedy, accurate, and accessible information from a trusted source, but many gave examples of poor communication [FG1, 5–6, 8, 13, I1–2]. For example, one participant said,

> local council got in touch with me and said, 'We are going to evacuate you' when the flood warnings came, we had that every so often on the phone, and you think, oh crumbs, it's getting serious now. And then the council said, 'Get your stuff ready, we're taking you out, we're moving you and the animals…but I never got evacuated [FG8].

Another participant said that having better communication and knowing what's actually happening would certainly make me sleep better when it's tipping it down with rain [FG13].

Most groups agreed that there was a need for better communication between all concerned parties when it comes to flooding, and this was something that their communities lacked based on their recent flood experience [FG1, 5–6, 8, 13, I1–2].

2.3.9.1 Risk Communication

Before the flood, a few participants felt that a *barrier to receiving communication* included the lack of awareness and action to inform the communities about their flood risk [FG2–3, 5, 12, I1]. One participant said he "wish there was some kind of campaign [to inform us]," and another in the group said the officials should "start doing a flood fair and telling people [about flooding]" [FG2]. In a different group, one participant said, "we didn't get clear guidance as to the extent of the flooding until it had virtually happened" [FG12]. Another participant stated the barrier as "information, I don't think, is always accessible as to whether these homes are at risk of flooding" [FG5]. Regarding warning systems for floods, one participant said, "we didn't really have much warning of this happening in 2013" [FG12], which he considered a huge barrier. Another participant in the same group said,

> we hadn't registered with the Flood Line because we'd never flooded before, and we didn't know we were at risk of flooding because nobody had ever flooded before, so we wouldn't think to register [FG12].

Knowing "how the flood is managed and perhaps having more information beforehand" were suggestions from participants on communication they wanted to receive [FG2, 13, I1].

Before the flood, many groups said there was a *lack of information from the local authority, officials, and outside organizations* and poor communication between them [FG2–3, 7–9, 15, I1]. For example, one participant described the issue of drains and sewers in the community. She said, “the normal drains are Surrey County Council [and] the sewers are Thames Water, and the two just never get together” [FG14]. Another issue brought up by the groups was the content of the risk messages by local authority and officials. One participant said,

> it’s just giving people enough information, not panicking them but giving them realistic, accurate information, and I don’t think they know how to do that [FG12].

Another participant said he heard afterward that there were:

> Very, very accurate warnings, but in 2013, we didn’t get those warnings. We had just the standard flood warning, which is a lower level of warning, and yet it came up much higher, and no one was coming around. So they need to get on top of that and to give us that information when we need to start moving stuff up[stairs] [FG12].

In a different group, a participant said he “get people [officials] coming and knocking on the door to say this [massive flooding] is going to happen [today] and then it doesn’t [happen]” [FG13]. Flood maps by the EA were discussed as well. One participant said, “Environment Agency maps don’t actually map what happens [in the communities]” [FG5]. Another participant said she printed the maps and went to a public forum organized by the EA to tell them about the inaccuracy of the maps with hopes to change them, but nothing happened to her efforts [FG3]. Many participants had conflicting and opposing views on the usefulness of the telephone service for flood information. One participant said her local authorities

> always, always took a phone call, and they were polite, and they were saying, ‘we will get to you as soon as we can,’ and that was some kind of reassurance really [FG1].

Another participant said the automated phone calls that he received from the EA made him “feel scared” and “that’s not very nice for elderly people” [FG3]. He did agree that the calls could be helpful if it was a real person instead of a robotic voice sharing the emergency alert [FG3].

The information from within or between communities before the flood was helpful to many participants, as it overlapped with the elements of preparedness and local knowledge. One participant said, “certainly the various stories that have been told to us about the flooding” helped her and her family stay informed about the flood risks [FG1]. Another participant shared his experience with his community’s efforts to educate his neighbors about the flooding risk. They passed out “a little leaflet” that, “said ‘look outside your front door and know where the drains are in your area’” [FG2]. He thought that helped to increase his neighbor’s awareness [FG2]. Other groups shared similar stories about informing their neighbors about flood risks—from going to “various houses and shops and a pub” and discussing in-person to using social networking sites like Facebook [FG1–3, 5, 12, I1]. One participant said the key to “keep the awareness up” in her community was having “things happening all the time” about the flood risks and preparedness along with acting upon lessons identified from prior floods [FG2]. “A lot more information just in the village about what to do, [and] where you can go for advice” was said to either be available or needed to be for the future in many of the participant’s communities [FG2–3, 5–6, 9, 12, I1].

Before the flood, many groups said there was *strong mistrust of the communication about flood risks.* One participant said this mistrust was because it was an "out of sight, out of mind" mentality [FG2]. Other groups shared similar notions saying people did not want to accept the flood risks and preparedness. One participant said it took "two floods for people to do it [in his community]" [FG2]. Another participant shared a story of someone who went with pictures from the past flood to new homeowners in the community to inform them about their flood risks [FG3, I1]. The new homeowners said this risk was mistaken because their estate agent would have told them about the flood risks when buying the home [FG3, I1]. In a different group, a participant said a person showed him and his family photos from previous floods, and if they had "not spoken to him and forewarned us, we would've had no idea at all" [FG10]. A few participants said their neighbors thought they were "mad" when they tried to inform them about the flood risks [FG2–3, 5, I1]. For example, one participant shared a conversation with her neighbors,

> also, people who expected it to flood the following Christmas – I said, 'No, it can flood any time of the year, it's not to do with Christmas, it's to do with the weather, and I tried to explain that, and people again thought I was completely mad [FG2].

2.3.9.2 Crisis Communication

Some participants felt a *barrier to receiving communication* during and after the flood. This overlapped with the element of community networks and relationships. One participant described it as, "you can't get out the house, you've got no email, computer facility, and you can't speak to anybody" [FG2]. She described how she felt isolated not only by the physical water barrier which the flood created but also by not receiving any pertinent information from her neighbors and local officials. Another participant suggested that

> whatever resilience plan you've got in place; you have to bear in mind that there are people who can't communicate, and you probably can't even communicate with anybody by knocking on the door cause you can't get to it [FG2].

Participants either witnessed or experienced how the flood waters can physically isolate homes from their community. To reach such houses to check in with them was difficult to accomplish.

Participants also discussed *the crisis information they received* from the local authority, government officials, and other organizations outside their community. Several participants shared their experience of receiving varied, at times conflicting, information during and after the flood. One participant described it as "different answers from the Environment Agency, the council; that was the problem, people thinking well, what do you believe?" [FG15]. Another participant described a message she received as,

> Environment Agency left a message on my voicemail. I can't remember exactly what it said, but it said that it will become dangerous, please evacuate, but that actually wasn't true information. So, if you weren't quite aware of what was going on that's quite scary [FG7].

Another participant described his experience as,

> we were standing in water where it was over our knees outside our front door, and the Environment Agency said quite categorically, 'No, there is no flooding in your specific area' [FG9].

Some participants spoke about difficulties accessing information and a lack of clear plans for action within messages they received. One participant said the officials had "no following through what they've actually said they were going to do" [FG10]. While discussing the issues of clogged drains in the community after the flood, many participants in one group agreed that this problem was because of a lack of communication between Surrey County Council and Thames Water in fixing this issue [FG14]. Another participant described the lack of accessibility of the information disseminated during the flooding. The "information that the local authority put out was not easily accessible" [FG6]. Another participant noted that he received little written advice months after the flood in his community: "at that time, February 2014, there was very little prepared written advice, documented by the local authority" [FG6].

According to most participants, *crisis communication given from within a community* was better with helpful content and accessibility than any information provided from outside the community. One participant thanked her neighbor who was in the group for keeping everyone informed:

> Communication was a great help to us anyway, having you living next door, knowing that we could ask you anything and you'd have the right answer. This was the thing so important. [FG11].

In another group, a participant said he was "very reliant on the community to be able to find out what's going on" [FG13]. "Most of our communication was amongst us as neighbors," said another participant [FG3]. She went on to say that this communication stemmed from a collective feeling of "oh my god, it's getting higher, we need to do something about it'" [FG3]. Similarly, a participant in a different group found this collective feeling as the reason for better communication seen locally:

> People are passionate about it because it's affecting them so badly rather than just people that sit behind a desk that have probably never had to deal with a flood situation in real life [FG7].

Overall, one participant summed up the powerful impact of what she saw from the community's communication to those affected:

> We were providing, I'd say, care and reassurance more than anything for people because they didn't know where to get it from, and they were panic – and just for us to sit down with people and say, 'We're here, here's our phone number. Don't worry. Do you need us to come and look at your property?' [FG6].

Many groups complemented social networking sites, such as a Facebook community page, which helped disseminate information within the community [FG6, 7 8, 11, I2]. One participant said the "social networking was very, very good" and said his community's Facebook page had "a massively positive effect" [FG6]. Another participant said her community's "Facebook page will make us more resilient because we can communicate with one another instantly" [FG8].

Concerns within communities about communication during and after the flood were tied to the concerns about the necessary action needed to resolve unexpected difficulties caused by the flood. For example, one participant said:

> One of the main concerns of all the neighbors down there was that the people who were older and who were on their own if they needed help, they got it, that was the primary concern of all the neighbors down there, and we were aware that there were people up here as well and we were trying to do what we could there, but we couldn't always get up the road to be honest [FG7].

This participant's concern was making sure help reached the appropriate people, such as the vulnerable people in his community.

Several groups in Surrey and Kent discussed *the intrusion of the media reporting and the general public visiting* the flood-affected communities during and after the flood. This created what a few participants called "a peep show," "a showground," or "a national playground" in their community [FG5, 9–10, I1]. The media and the general public acted more as tourists to see the flood-affected communities than to help them, mainly seen negatively by the groups [FG5, 9–10, I1]. One participant elaborated on the "peep show" analogy:

> We were a peep show. To be honest, we were a peep show. They closed the road down because of these pipes going across, and daily and weekends especially, I felt like I was on show. And one gentleman, actually we were gathered on the bridge talking with neighbors, and this gentleman walked along, and he walked down our pathway, and he lent over our flood barrier door and was taking photographs [FG10].

In a different group, three participants saw this intrusion as a lack of responsible journalism:

> I mentioned insensitive media; it worked totally against us. It didn't work for us; it didn't work with us; it worked against us. It was just sensation hunting. They just wanted to see people in distress [FG9].
>
> The way they worked the area was disastrous, and in fact, I was quite happy to drag them out by their ear and kick them out [FG9].
>
> They were dangerous, they were causing more problems, and they got in the way. And it was irritating, these people just kept turning up on your doorstep, and you just wanted to try and get on with your life. [FG9].

However, some participants, especially during the first few days of the flood, felt the media coverage was good for their community because of the information, attention, and awareness it brought to their community. All Surrey-based groups noted that the media came before any help from emergency services [FG7–9, 11–12, 14–15]:

> TV vans, cameras, and reporters were here well before any emergency services [FG9].
>
> The press were here on the Monday night before everybody else arrived [FG11].

Because of the media's quick entrance to the Surrey regions, a participant felt that the services were here much quicker because they could probably show off in front of the cameras that they are doing something, and they were doing it in the area which was already well known to the TV crews [FG9].

Similarly, in a different Surrey-based focus group, a participant said,

Sky News came, and then they started filming on our road, and that's when we actually started to get some help, but before that, we were just really left to our own devices, and nobody seemed to care about us or would tell us anything [FG12].

Few had positive things to say about the media in Kent-based groups, but one participant credited the radio and television as excellent and praised Kent Radio as "very good" [FG1]. Radio was the preferred choice for receiving news during and after the flood in most Kent-based groups [FG1–5, I1].

2.3.10 Governance and Leadership

Discussions on governance and leadership had many participants feeling frustrated and angry about their experiences. Key topics included *having better leaders, the blame game type of politics*, and *the lack of responsibilities by services and officials*. Additionally, the groups discussed *the services and officials not being prepared or able to help communities, community members giving crisis information to help with services, the work by the tankers and workers*, and *the lack of public involvement and support*.

Many groups discussed *having better leaders and the qualities of a good leader* who needed to oversee governance and post-flood services in the community. One participant said,

one of the biggest things that I would have to say is around the whole governance and leadership part. I think leadership is something that it isn't just about somebody leading. It's not just a practical requirement of having a leader. It's picking the right leader. It's having somebody who will listen, who understands, who has common sense, who's a good communicator, who is inspirational, who is somebody positive, somebody, who can deal with negative situations, somebody who can calm people [FG11].

Another participant in the same group discussed her observation of a good leader in her community after the flood:

His leadership was very much of a team leader, and we [she and her community] all felt very much as important as he did, and that's really important to a community, to make everybody feel that their role, no matter whether it's redirecting traffic or serving cups of tea, it's all of the same value [FG11].

In a different group, one participant said,

two or three members of the parish council really took it upon themselves to actually take responsibility for certain aspects of it [leadership and coordination], which I think was quite important [FG1].

Tied with better leaders was also the issue with the building infrastructure in the communities. One participant said,

> the neglect of the infrastructure and also of the constant building on the flood plain, which continues now, is a major one, and it's something that's not mentioned very much because it's inconvenient, but it's the truth [FG8].

However, another participant in the same group suspected that "there's the political aspect of this which is that a lot of very central infrastructure is in private ownership" that can pay a lot of money to get their building accomplished quickly, which the local authorities are looking for, such as with housing and retail development [FG8].

Participants also discussed *the lack of responsibilities by services and officials* and *the blame game type of politics* they saw after the flood. Participants said,

> each governing body won't accept responsibility. They blame the other, so the EA will say it's Southern Water's fault, Southern Water will say it's EA's fault, the council will say it's somebody else's fault [FG10].

Many participants felt that *services and officials were not prepared or able to help their community*. One participant said, "because we're in a community, not one of the large ones [that were flooded], we were just left on our own" [FG10]. Another participant said a "terrible shortage of emergency services because generally, they were all on skeleton staff." In the same group, a different participant said, "it was almost impossible to get any of the emergency services to do anything" [FG2]. In terms of being able to help her community, a participant said,

> I think my experience from the word go was there was no leadership because there was no real understanding of what to do by the people in that meeting, in COBRA meeting or in the Silver meeting or in the Bronze meeting, they're in a room, and that was the biggest problem that they have [FG11].

Further touching on these Cobra meetings, which were the crisis command meetings organized by the government, a participant in a different group said,

> there was certainly the gold, silver, bronze command level, but the reality is, at the absolute sharp end of who is going to help a resident that is either flooded or potentially could be flooded, would be the local community, community groups and so forth [FG6].

Because of this, one participant discussed why he thought services and officials were not prepared and said,

> [the] big unknown is whether the services and the councils and local governments will rely on 'no, this can't happen again for a long time because floods like this don't happen very often,' but they do [FG9].

Participants also discussed how *community members were giving crisis information to help with services* they needed but were not receiving. Many participants echoed frustration along those lines. Their communities "needed proper organized help" [FG10] or at least "practical help" to cope with the flood [FG5]. Several participants had similar discussions to the participant who said,

> it [the help from officials] was a bit late. You could've done with some of that quite a bit earlier on in the process. I think a lot of people just weren't able to cope and had to admit that, and they couldn't turn to anywhere else for help and had to move out [FG12].

Most groups mentioned *the work of the tankers and workers*, especially with the pumping of water and sewage. During and after the flood, these workers provided a vital service during and after the flood, and many participants found their service helpful [FG1, 3, 5, 10, I1]. Another participant said she saw the workers with tankers "from the window here at night, and they were coming out every night, they were pumping," and she called their service "excellent" [FG1]. Similarly, a participant said his "positive experience was probably, yes, it was the two men that were on the ground that helped us" [FG10].

Participants wanted *more public involvement and support* in the leadership and governance after the flood. Many participants felt they were not listened to, and many thought that the leadership should have acquired some local knowledge, as mentioned in the elements of communication and local knowledge. Additionally, a participant said the ownership and responsibility for rebuilding life after the flood should be "back to the locals who could actually tell them, 'Hey, this is what we need'" [FG10]. He went on to say that there should be more "self-guided, self-help from the community" with the efforts after the flood [FG10]. In a different group, one participant discussed how she was not being listened to about how building more properties on flood risk areas would harm the community:

> After the flood, the council is still building properties in flood risk areas, which isn't helping the whole problem with flooding, and it's going to cause more financial cost to everybody really [FG12].

Overall, participants were dissatisfied with the quality of governance and leadership in response to the flooding. Many made negative comments and expressed emotions which bordered on feelings of neglect. Unfortunately, there have been similar situations where disaster has struck that has resulted in similar attitudes. For example, in reponse to the Flint Water crisis in the United States, many media outlets accused the ex-Michigan governor for neglectful behaviour [7]. In addition, much has been said about disasters such as Hurricane Katrina in the United States and Fukushima nuclear disaster in Japan that both the physical and psychological consequences could have been reduced if it had not been for poor governance and leadership [18, 45].

2.3.11 Mental Outlook

Nine key topics emerged from the focus group discussions around the mental outlook. They included *adapting to a post-flood community, positive and negative coping experiences, previous flood experiences, the sudden shock of the flood*, and *trauma caused by the flood*. Many participants also *felt worried sick and did not have any trust or faith*, and even developed a perspective of *a long-lasting fear for the future flood* and *a strong belief of never being able to recover and not having any hope for the future*. Additionally, many groups discussed *the complacency for things to change* in the community after the flood.

Adapting to a post-flood community was quickly brought up by the participants and explicitly entailed how they had to adjust to life after the flood [FG1–3, 5–9, 11–15, I1–2]. Many participants described both poor and suitable coping mechanisms they used after the flood, while others struggled to adapt to life after the flood. Participants attributed this struggle to the inability to mentally move forward from what they underwent in the flood experience, which included the overlapping impact of the elements of economic investment, community networks and relationships and resources (i.e., the strain of flood insurance, lack of local support and resources).

Adapting to life after the flood was described with mixed feelings. Some participants described poor adaptability to the uncertainty of life after the flood, such as having "the 'poor old me' syndrome" [FG10] or recalling the experience as "very, very depressing" [FG15]. Other participants said they had better adaptability to life after the flood, as many recalled that they had to re-focus themselves to move on with their life by any means necessary [FG10]. One participant described the difference and his choice on how to adapt to life after the flood in the following quote:

> And it wasn't poor old me. We [he and his community] just got on with the job; we knew people weren't there to help us, so we got on with it ourselves, but sympathy was given to people who did get flooded, but they did the poor old me syndrome. They didn't do anything to prevent themselves from flooding, but they got the sympathy, I think more [FG10].

Participants took further steps to re-focus themselves and described these mechanisms as part of *positive and negative coping strategies*. Positively, participants adopted the 'blitz spirit.' One participant said:

> I felt that we adopted an in-the-war attitude towards it all. The atmosphere [during and after the flood] was like we were in the, you wouldn't know, like we were in the blitz, fighting spirit [FG15].

Another participant in a different group recalled the same feeling and described it as "like sort of blitz spirit replaced by desolation and desperation" [FG9]. Blitz spirit, discussed in three groups, was seen as a positive coping mechanism for the uncertainty of life after the flood [FG3, 9, 15]. Another positive coping strategy used by the participants was mentally focusing themselves on carrying on and move forward:

> There's nothing more sure than tomorrow is going to come, the clock, the time's going to keep ticking on, and yes, people go back, they have to, we have to, we have to go back to work, we have to carry on our lives [FG11].
>
> We're past the shock and horror and anger, and into let's just get on with it because we all still want to live here [FG2].
>
> Try and just keep going and deal with what was happening [FG5].

This idea helped many participants to carry on with their daily life quickly. One participant described how his family was able to adapt to the flood impacts quickly:

We managed to clear up within two or three days enough and carry on with normal life. Kids went to school, I went to work, and everything happened, but with all our stuff out in the street. So, in that regard, it sort of carried on quite normally [FG13].

Another participant said being active and valuable to her community helped her during the flood:

While we were being flooded, it was fine because we were all doing stuff, we were sorting food for people, we were doing things, and we could do something [FG9].

Negative coping strategies were used and described by the participants as well. Several participants felt they were coping poorly and found life after the flood "much harder to cope with" [FG9]. Another participant discussed her difficulty dealing with losing irreplaceable items in her life:

Photographs and personal items that were ruined are irreplaceable, so that's difficult to deal with, and we [her community] have to have coping mechanisms in place to be able to deal with it [FG2].

Other participants said they took a long time to adapt to post-flood life and even suggested that they may not ever cope with their flooding experience [FG5–6, 9–10, I1]:

It's not an overnight thing; it takes time - it takes years; you've got to get over it [FG10].

Recovery is not instant or overnight. It could be years. In fact, in the absolutely worst case, it could be never [FG6].

Trying to cope with such eventualities, there was this realization that it's not going to go away [FG5].

Overall, participants understood that time was needed for recovery. Still, their expectations of being helped right away were not met, which delayed their "healing process" [FG10].

A few participants discussed their *previous flood experience* and how it matched up with the most recent flood. One participant said, "it was less stressful the second time around. It was very stressful the first time when you found water in your house" [FG3]. Another participant found the "second flood much less stressful because of knowledge gained at first flood" [FG3]. However, the opposite feeling was also seen and discussed by participants. Some participants suggest that their experience of the subsequent flood was worse than that of the previous flooding. "Not nearly as bad the second [flood]," said one participant when asked about her first flood experience [FG3]. Another participant said he and his family bounced back quickly in a 2001 flood but found the recent flood leaving them with a "very bad" feeling [FG5].

The sudden shock of the flooding event affected the mental outlook of the participants. One participant compared the flooding event to a horror movie. She said that she saw

this water was coming, not actually through the front door, but coming through the skirtings, coming through, and it was just like a horror movie. I just couldn't believe it [FG13].

Another participant described her progression of feelings through the event:

> In the beginning, when it first happened, the first couple of days there was a mild panic, I suppose, and fear; but as the days progressed into the third and fourth day, there definitely was a different feeling. I think fear had gone, and it did become a real depression. And certainly, in here you could feel, you could just feel it [FG15].

A similar feeling was described by another participant when talking about life after the flood waters went away:

> Once the water had gone, that was when it really hit you, and you just looked at the mess and the destruction and the smell, and your heart just sunk really. Because you just thought right, now I've got to deal with this [FG9].

Furthermore, the shock and flooding also created sudden isolation, which led to fear, according to a participant:

> The village was cut off for about an hour, an hour-and-a-half, and there were hysterics by residents who wanted to get out to their mothers, fathers, children, whatever [FG1].

Regardless of how adaptive the participants felt after the flood, there was consensus across the groups on *the trauma caused by the flood*. Participants recalled the flood experience as a traumatic experience:

> Whether you ever recover from that and are the same person you are afterward as you were before, I think you have to view it as a trauma. It's a traumatic experience [FG11].
>
> High percentage of the village, of this community, have suffered, some are suffering post-traumatic stress. It's shown in high emotion. It's shown in inappropriate outbursts. It's shown in people just not wanting to talk [FG11].
>
> I was absolutely at the end of my tether looking after my mother with dementia, and I've got four children, and I didn't have the resilience to deal with the flood [FG3].

Trauma from the flood was compared to trauma felt from a burglary from one participant:

> I suppose it's very much like a burglary. Once someone's been into your house and got into your privacy, you feel that everything feels unclean, and it's the same with the floodwater. It didn't come in the house, but we came in with wet boots, we came in with leggings and everything [FG11].

One participant in a different group stated how she could not sleep after the flood [FG5]. Another participant in that group described life afterward as "mentally exhausting because you couldn't concentrate on anything else all day" [FG5].

Feelings of *being worried sick and not having any trust or faith* were felt by several participants during and after the flood. Participants described the feelings as "just a truly, truly horrific experience" [FG2] and "you're frightened and vulnerable and leaving your home as well is almost more than you can bear" [FG2]. In a different group, one participant explained his feelings as a loss of trust because he was distraught with his local officials and lack of support during the flood [FG9]. Similarly, another participant said she "lost the faith," and even though she survived and moved on, "it is that impact; you don't have the trust anymore" [FG10]. This "impact" was also seen in being worried sick about the ongoing flood progression and its impact on the community:

> You're worried that you're going to lose everything, and what little bit you have got and saved for is going to end up underwater, and that's very frightening [FG14].

Furthermore, several participants in different groups remembered: "crying," "worrying," and getting "quite depressed about the whole thing" [FG3, 8, 9, 14, 15]. One participant did her best not to shed a tear when sharing her feelings and experiences with the group, but she said, "there were a lot of tears, very much so" [FG9] during and after the flood.

Within three groups, participants felt a division in the community between those directly affected by flooding and those who were not [FG3, 5, 7]. This created a loss of trust in the community. One participant felt a total sense of "antagonism, totally unnecessary, and unkind antagonism" from surrounding non-flooded neighbors in her community [FG5]. According to the participants, this division stemmed from the loss of trust and faith in the neighbors to help them through the flood [FG3, 5, 7].

Almost all participants suggested *the outcome of long-lasting fear* based on their flood experiences. Participants said:

> There was a huge underlying anxiety [FG5].
>
> Things after the flood have been way more stressful than the actual flood itself, I feel [FG7].

Many participants left being fearful of the UK's rain and the winter season when remembering these feelings:

> I feel like I'm treading on broken glass waiting for it, and when it gets to winter time and autumn and it starts that heavy raining, like this last year, every day I'm going and looking over the bridge [to see if it will be flooded] [FG10].

One participant in a different group explained this long-lasting fear as,

> if you've been flooded, you live with it all the time, and you get - obsessive is the wrong word, but one worries about it [FG2].

Beyond this, some participants felt they would *never recover from the flood and had no hope for a better future* [FG3, 6, 10, I1]. One participant used the idiom of "canary in a coal mine" to describe his depressing view of the future, stating that he is "still going to be the canary in the mine, and that's a bit depressing" [FG3]. Another participant strongly believed that in the future, her community would continue to be "allowed" to flood as a useful early indicator to prevent other neighborhoods closer to London from flooding and, in essence, "to make London more resilient" [FG6].

Many participants saw that *complacency set in after the flood*, which increased their fear and frustration in life afterward [FG1, 3, 10, 12 I1]. For example, one participant stated her frustration about her community's future:

> Complacency has set in, that's the problem, and nothing's been done, and we know it's going to happen again, and we know that there's the opportunity to actually change things, and that's not being taken - so that's really frustrating [FG12].

Another participant described the situation as "we know that it's likely to happen again, and rather sooner than later, as things are going to get wetter rather than dryer"

[FG1]. Still, according to these participants, the actions and progress in developing preventative measures for the next flood in their community have yet to be prioritized [FG1, 3, 10, 12 I1].

2.3.12 *Matching Key Topics from Focus Groups to the Taxonomy of Community Resilience Elements*

Table 2.6 shows the key topics identified in the focus groups and how these map onto the taxonomy of community resilience elements identified in Patel et al. [37] and discussed in Chap. 1. Some of the key topics found in the focus group discussions were reorganized from where they emerged in the conversation about an element to map with the identified taxonomy. For example, the focus group participants identified the dangers of flooding, flood water, and sewage during the discussion on the element of health. However, when considering the descriptions of the elements of community resilience found, this key topic aligned better with the sub-element of training and education within the aspect of local knowledge. This realignment allowed a better union between scientific literature and qualitative findings.

Examining both columns in Table 2.6, all elements of community resilience were present in both findings. None of the critical topics led to needing a new element. Only one new sub-element, pre- and during disaster economic plans, needed to be formed within the element of economic investment. Overall, the key topics from the focus groups provided additional details for each component.

2.4 Discussion

Participants from the focus groups shared their insights and experiences from dealing with flooding in their community. The discussions helped provide evidence as to whether the taxonomy of community resilience elements could be applied to UK communities that went through flooding. In essence, the findings of Patel et al. [37] were supported by the results of this qualitative research. Still, many of the details for these elements of community resilience were better explained in terms of how they helped or could have helped people reduce the harmful security and health impact caused by the flood. Some sub-elements had more examples than others, and a new sub-element was found within the element of economic investment. The findings from the focus group discussions supported the practical usefulness of applying community resilience elements.

Similar to the systematic review findings by Patel et al. [37], a common meaning of community resilience was not found in the focus group discussions. The definitions used in the communities were more as a catch-all term for reducing harm against flooding than as a specific, clearly defined concept.

Table 2.6 Matching key topics from focus groups to the elements found

Community resilience elements	Sub-elements	Key topics from focus groups
Local knowledge	Factual knowledge base	Taking actions on prior knowledge Knowing who to call or ask Community's history of rivers and floods Community awareness of flood risk
	Collective efficacy and empowerment	Community members buying or building own resources Community members helping each other
	Training and education	Dangers of flooding, flood water, and sewage
Community networks and relationships	Connectedness	Community connectedness Outside organization helping the community
	Cohesion	Community spirit Isolation and abandonment
Communication	Risk communication	Barriers to receive communication Information from authority outside Information from between community Mistrust of communication
	Crisis communication	Barriers to receive communication Information from authority outside Information from between community Mistrust of communication
	Effective communication	Effective communication Impact of media, filming crews, and tourists
Health	Health services	Quality and delivery of health services
	Physical health	Issues with physical health
	Mental health	Problems with mental health

(continued)

Table 2.6 (continued)

Community resilience elements	Sub-elements	Key topics from focus groups
Governance and leadership	Infrastructure and services	Sewage and clean water concerns Tankers and workers Action to ameliorate the sewage issues Having better leaders Blame game type of politics and lack of responsibilities by services and officials
	Public involvement and support	Community members giving crisis information to help with services Services and officials not being prepared or able to help communities Lack of public involvement and support
Resources	Resources	Wrong type or lack of resources Resources and essential goods bought by communities Leftover resources and post-flood improvements Resources that were said to help Sandbags
Economic Investment	Post-disaster economic development	Costs after the flood Post-disaster spending
	Post-disaster economic programming	Government grants and post-disaster economic program Insurance companies and the strain of insurance
	Pre- and during disaster economic plan*	Spending to prepare for a flood Spending during the flood
Preparedness	Planning and mitigation	Lack of preparedness and plans Grassroots efforts by the communities Physical flood defenses Lessons identified and outcomes for action Future needs to improve the community's readiness and crisis management

(continued)

Table 2.6 (continued)

Community resilience elements	Sub-elements	Key topics from focus groups
Mental Outlook	Hope	Feelings of worried sick and not having any trust or faith Belief in never being able to recover and not having any hope for the future Complacency for things to change after the flood Outcome of long-lasting fear
	Adaptability	Adapting to a post-flood community Positive and negative coping experiences Previous flood experiences Sudden shock of the flood Trauma caused by the flood

*New item to the previous taxonomy

However, many of the participants' responses did align with previously used terminology and even a few general types of definitions. Many participants interpreted community resilience as the ability to "bounce back" after the flood. This was similar to the descriptions by Paton et al. [40, p. 158] ("...community characteristics and processes that promote a capability to 'bounce back'..."), Twigg [53, p. 6] ("...capacity to recover or 'bounce back' after an event"), Cox and Perry [14, p. 396] ("...the capability of a community to face a threat, survive and bounce back...") and Lowe et al. [29, p. 2] ("... the ability to "bounce back" from disaster"). The participants also gave definitions with a range of "planning, protocols, and support" attributes. This was similar to the general type of 'range of positive attributes' definition (i.e., a broad collection of response-related abilities) found in publications by the UK Cabinet [8, p. 11] and Coles and Buckle [12, p. 6]). A few participants even gave 'process' definitions (i.e., an ongoing process of change and adaptation), as seen in publications by Lemyre et al. [28, p. 319] and Norris et al. [35, p. 131]. A blend of these was also seen in the focus group discussions, which aligned with the definition by Pfefferbaum et al. [41, pp. 241–242]. However, the participants understood the core elements of community resilience derived from Patel et al. [37] and provided examples and anecdotes on how it applied to them based on their past flood experiences.

Furthermore, the discussion on the elements of community resilience started before the participants were prompted to discuss them. The participants gave examples of these elements as things that helped them with their flood experience. This clearly showed the wealth of knowledge and insight that the participants gained from their flooding experience and their journey toward reducing the flood impacts. By directly asking what a community resilient to flooding means to them, participants described what most definitions did: a catch-all term but based on first-hand evidence. A few participants discussed community resilience as it applies to flooding as firm beliefs of either 'bouncing back' or 'a process of change.' Most focused on solutions that would help or push their community better suited against flooding. For example, one participant said the notion of bouncing back "sometimes means the lessons haven't been learned and you're maybe not as prepared as you might be" [FG11]. To avoid this, below is a summary of what was learned for each element (denoted in *italics*) from the focus groups.

2.4.1 Summary of Elements Found in Focus Groups Discussing Flooding

The focus group discussions showed that some aspects of community resilience had a more significant impact on the communities than others in reducing the harm caused by flooding. These elements tended to vary in strength against reducing damage depending on if it occurred before, during, or after the flood. Before the flood, participants said preparedness and aspects of communication, health, economic investment, local knowledge, and governance and leadership impacted them. Participants said the most impactful elements were resources and community networks and relationships during the flood. After the flood, mental outlook, primarily along with all the other elements, were more impactful. This adds a new understanding of community resilience: its taxonomy changes depending on when applied to flooding.

Local knowledge was explained as the concrete information, education, and experience acquired from being flooded. Many participants expressed a feeling of being confident in knowing who to call or ask, which empowered people to get involved and help one another. Most participants valued learning the history of floods in their community. Some even created annual events as a remembrance and an opportunity to increase the awareness of their community's flood risk.

Community networks and relationships were defined as the connections in and out of the community and the nature of these connections. Almost all participants understood and expressed the importance of their relationships within their community. This contributed to the idea of community members helping each other as an essential part of a participant's ability to recover from the floods. Getting to know the community members and strengthening the ties among neighbors may be a priority for people in flood-risk areas to reduce feelings of isolation and abandonment, especially if no immediate outside aid comes after a flood.

Participants in the focus groups said they received poor risk and crisis *communication*. The great lengths that must be made for people to accept that they may be flooded could be remedied by having effective flood-related communication occurring in the community. As a result, the groups highlighted the many barriers to receiving flood-related communication and expressed a growing mistrust of the local authority's risk communication within the community. Overcoming the lack of trust and confidence with the official risk communication sent into the community will take time to rebuild, as many participants remembered how distraught they felt from their prior flood experience.

The community's *health* was a concern for many participants, especially in terms of the trauma caused by the floods. Multiple participants reported that the health services serving the community were overwhelmed, but good quality help was given to them once they arrived in the community. Essential services should not make a community feel that their health is unimportant. A more considerable emphasis on mental health in the community is also required from emergency workers when helping to provide long-term follow-up care.

According to participants, a lack of *leadership and governance* was seen in the public services and infrastructure post-floods. Many participants voiced their concerns that their prior experience with governance did not instill confidence or trust in the leadership to handle a future flood well. Preparing current leaders in and out of communities with better flood management skills were lessons that many participants said they learned and hoped their government and authorities did too. Drawing from the elements of resources and preparedness, many groups gave examples of things that were much needed and should be considered at all levels of government.

Most *resources*, besides sandbags, were said to be provided by the community members rather than by authorities. Community resilience included neighbors making market runs for people and supermarkets donating food to communities. Tangible resources such as portaloos were discussed as being much needed immediately after the flood by the participants. Also, a concern from many participants that may need further attention was how to make resources used against the flood reusable for the community. For example, the sandbags could be reusable if there was storage in the community where people could drop and pick them up from.

The element of *economic investment* seemed to be disenchanting for many participants. Most participants were frustrated with the reimbursement process of insurance companies and the complicated process of claiming government grants. According to a few participants, these grants were "secretive" and "hard for even the educated to fill out." Many participants did not know about these grants despite their apparent eligibility for receiving them.

From employing flood barriers and sandbags, most of the *preparedness* activities mentioned by the groups revolved around learning lessons and getting things right for the next flood. As one interviewee said, the issue with preparedness in the communities was that people were more inclined to react than a plan for flooding. Many participants said, most community members have a short-term or no memory of flooding. This held especially true among participants who said their homes were not affected by past floods.

The participant's views and feelings on *mental outlook* as it applies to flooding were especially powerful. Participants intimately shared their deep emotions, and many were leaning towards post-traumatic growth or long-lasting fear after the disaster. The effects of the trauma caused by the flood and the uncertainty of the future were reported to be overwhelming and scary by many groups.

2.4.2 Associations Between NRR Security Outcomes and Elements of Community Resilience Described in Focus Groups

Suppose elements of community resilience could accurately predict the level of security and health impact a community will face from a future flood, then building prevention in communities before a flood could reduce any likely harm impact. From the evidence in this Chapter, most participants were keen to know how they could reduce the harmful security and health outcomes caused by flooding, which are also described by the UK government as the NRR security outcomes (psychological impact, economic harm, and social disruption). While working through examples and anecdotes on elements of community resilience, participants would point out how their examples were a benefit or harm caused by the flood. This shows that an association between an element and an NRR security outcome(s) could be possible and valuable to inform the preparation for future floods. For example, many participants described the importance of having a cohesive community, especially with their immediate neighbors. Participants saw this cohesiveness being built strongest right after the recent flood. In a study by Chang [11] community cohesion increased immediately after the flood but reduced long after the initial onset of the flood. Many participants in the focus group felt this reduction in cohesiveness and suggested investing efforts to keep community members lively and engaged in preparing for the next flood.

This Chapter's results align with the current understanding from prior studies on communities affected by flooding. For instance, the participants who were aware that they lived in a flood-prone area and those who were recently affected by the flood mentioned that they were acting to be better prepared for the next flood. This matched the findings found by Coulston and Deeny [13]. This author found that both flood risk awareness and the recent flood experience of the UK 2007 summer floods were motivational factors for their participants to better prepare for a future flood. In another example, many participants felt that better *local knowledge* and *preparedness* before a flood would have given them a faster recovery after the flood. This aligned with Thieken et al. [52] who found that people with better knowledge of self-protection measures and applying them were more advantageous for fast recovery after the flood than those who did not.

Findings from the focus groups also confirm and add to previous results that better *community networks and relationships* reduce harmful security and health outcomes caused by flooding [4, 5, 11, 20, 45]. Most groups discussed the *health* impacts caused by the floods, such as the heightened level of stress and trauma that occurred. This impact has been previously documented in flooding studies, which found flooding to lead to worse physical health [1, 27], likely PTSD [21, 32] and the onset of other mental health outcomes [1, 10, 23, 26, 30, 31, 33, 36, 44].

Of all the elements of community resilience, the participants strongly emphasized their *mental outlook*, standing out from what was previously found in the literature for this element's association with damage caused by flooding [3, 4, 25, 30, 34, 42, 48]. What stood out was that many participants saw the flood experience beyond a traumatic one and defined it as a new mental state for how they act and live moving forward in the future. Some participants saw themselves moving forward positively, which aligns with the publications on how disasters could create post-traumatic growth among those who were affected [39, 49–51]. Other participants in the focus groups saw themselves moving forward negatively, developing a long-lasting fear of rain and losing hope for a positive future. This viewpoint aligned with the results found by Werritty et al. [55] who looked at the flood impacts in Scotland between 1993 and 2005. Werritty et al. [55] listed long-lasting fear of floods as an "intangible lasting impact" caused by the flood. The authors posed a question on its long-term implication that people could move away from their communities if people's fear of future floods becomes greater. This may be true, but participants could also shift into a mentality closer to post-traumatic growth years later. Holgersen et al. [24] found that post-traumatic growth happening many years later in disaster survivors could reflect on how they managed their suffering. This could be true for the participants in certain focus groups, as this new mental state could result from how they managed their suffering. If so, then the element of mental outlook has a prominent role in applying community resilience.

2.4.3 Limitations

The research carried out in this chapter has limitations that should be considered. First, the elements of community resilience found in the focus groups discussion may not apply to other regions in the UK. The groups were from Kent and Surrey areas where flooding had occurred. There could be a disagreement with the elements or even a shift in priorities based on local, regional, and/or national perspectives. The participants were also British nationals, and the results could differ if carried out in other countries. The relative wealth of the country could be a factor here. These could also limit the generalizability of the results.

The participants were recruited based on their own self-defined status of being flooded. As a result, the discussions included people with varying degrees of the impact caused by the river, coastal, surface water, sewer, and groundwater flooding. Participants could have self-identified as being affected by flooding from being directly affected by flooding or indirectly affected by flooding, such as by having their route to work blocked. Also, the participants had a different understanding of what being flooded meant to them—from water entering their homes to water being a disruption outside their homes. Imposing a stricter definition of being flooded during the recruitment could have led to higher accuracy of having participants with similar flood status, but this may have reduced the number of people who could be recruited. Also, a stricter flood status check could have reinforced and contributed to the negative aspects of research fatigue among the population asked to join a post-disaster research study [38]. Regardless of the flood level or impact experienced by the participants, each village where the participants were from was checked if it was flooded and checked the village's flood risk using the EA's flood risk map. Each community also had a varying level of flood protection measures which differed and were not controlled. Future studies should examine if the different risk levels may have influenced the details received in this chapter. Despite the self-defined status and varying flood protection measures in the community, the goal of capturing information on the bigger picture of a flood-affected community was achieved.

The sample used in this Chapter was restricted in age and gender. Predominantly women older than 55 participated in the discussions. There could have been alternative viewpoints and/or exchanges on the elements if there had been a better balance in the age and gender in the groups. Also, the participants were flood experts—or at least people interested enough in the topic to attend a focus group. People with different levels of flood education might have had a different view on things.

Researcher bias may have affected the matching of the elements of community resilience and the key topics found in the focus group discussions. It is possible that one could have interpreted the key issues differently and seen how they fit into the elements of community resilience. For example, lessons identified and outcomes for action from the flood could have been categorized into local knowledge, governance and leadership, and preparedness. To limit this bias, the descriptions of the elements of community resilience were used to guide the analysis of difficult matches that were conferred with others on the project.

Lastly, because a framework-driven approach was taken and participants were provided with a booklet about this framework (i.e., elements of community resilience), the participants could have been prompted to give quotes and examples that the researcher was seeking. This approach was limited by asking for details before introducing the framework to probe for information about the elements of community resilience. Also, the participant's exposure to the framework was limited by instructing participants not to go ahead with the booklet until asked. Based on how the participants answered in the focus group discussions and written sections, most of the results would have still emerged if the framework had not been given. The benefit of using the framework by Patel et al. [37] was that it allowed the participants to discuss quickly and effectively the details that helped or harmed their flooding experience.

2.4.4 Implications

The results from this Chapter add to the growing list of implications that could be applied. For community leaders and emergency planners, the evidence provided in this Chapter adds further support to the scientific literature on the general elements needed to strengthen a community at-risk from a disaster such as flooding. Having discussions among officials, stakeholders, and leaders in a community to create realistic plans around each of the community resilience elements will allow communities to be prepared for a disaster risk and crisis and to reduce the likely harmful security and health impact, according to the conversations with the flooded-affected participants captured in this Chapter.

For colleagues and researchers using community resilience, the focus group discussions illuminated that the definitions of community resilience continue to vary in interpretation and implementation. When communicating community resilience to flood-affected communities, it may be best to specify precisely what is meant and to identify the specific element of interest. As seen in this Chapter, the elements of community resilience play a more prominent role in enhancing communities to reduce their security and health harm to disasters. Elements such as *community networks and relationships* and *mental outlook* were seen to be invaluable for many participants who experienced flooding and further work should investigate the benefits and associations with these elements to reduce likely harm created by a flood.

2.5 Conclusion

Previous work has identified community resilience elements as it applies to disasters but lacked the application to flooding in the UK. This Chapter aimed to understand this application to flooding in the UK via focus groups of people who have recently been affected by flooding. The aims were to understand the participant's perceptions of community resilience following a flood and examine the taxonomy of elements of community resilience as identified in the previous chapter and its relevance in reducing harm after the flood for the participants. Based on 15 focus groups and two interviews, totaling 77 flood-affected people in the UK, it was determined that the current list of elements does apply to severe flooding in the UK. However, the specific details on whether they work in reducing the harmful security and health outcomes caused by flooding are still unknown. The next Chapter will test the elements of community resilience and their likely association with reducing harmful flood impact among those living in high-risk flooding areas in the UK.

References

1. Alderman K, Turner LR, Tong S (2012) Floods and human health: a systematic review. Environ Int 47:37–47. https://doi.org/10.1016/j.envint.2012.06.003
2. Begg C, Walker G, Kuhlicke C (2015) Localism and flood risk management in England: the creation of new inequalities? Eviron Plann C Gov Policy 33(4):685–702. https://doi.org/10.1068/c12216
3. Bokszczanin A (2007) PTSD symptoms in children and adolescents 28 months after a flood: age and gender differences. J Trauma Stress 20(3):347–351. https://doi.org/10.1002/jts.20220
4. Bokszczanin A (2008) Parental support, family conflict, and overprotectiveness: predicting PTSD symptom levels of adolescents 28 months after a natural disaster. Anxiety Stress Coping 21(4):325–335. https://doi.org/10.1080/10615800801950584
5. Bokszczanin A (2012) Social support provided by adolescents following a disaster and perceived social support, sense of community at school, and proactive coping. Anxiety Stress Coping 25(5):575–592. https://doi.org/10.1080/10615806.2011.622374
6. Braun V, Clarke V (2006) Using thematic analysis in psychology. Qual Res Psychol 3(2):77–101. https://doi.org/10.1191/1478088706qp063oa
7. Brooks SK Patel SS (2021) Psychological consequences of the flint water crisis: a scoping review. Disaster Med Public Health Prep 1–11
8. Cabinet Office (2010) Draft strategic national framework on community resilience. London, UK. https://www.gov.uk/government/uploads/system/uploads/attachment_data/file/60922/Strategic-National-Framework-on-Community-Resilience_0.pdf
9. Cabinet Office (2015) National risk register for civil emergencies-2015 edition. Cabinet Office, London, UK. https://www.gov.uk/government/uploads/system/uploads/attachment_data/file/419549/20150331_2015-NRR-WA_Final.pdf
10. Chae EH, Tong Won K, Rhee SJ, Henderson TD (2005) The impact of flooding on the mental health of affected people in South Korea. Community Ment Health J 41(6):633–645. https://doi.org/10.1007/s10597-005-8845-6
11. Chang K (2010) Community cohesion after a natural disaster: insights from a Carlisle flood. Disasters 34(2):289–302. https://doi.org/10.1111/j.0361-3666.2009.01129.x
12. Coles E, Buckle P (2004). Developing community resilience as a foundation for effective disaster recovery. Aust J Emerg Manag 19(4):6. https://search.informit.com.au/documentSummary;dn=375435145094637;res=ielhss
13. Coulston JE, Deeny P (2010) Prior exposure to major flooding increases individual preparedness in high-risk populations. Prehosp Disaster Med 25(04):289–295. https://doi.org/10.1017/S1049023X00008219
14. Cox R, Perry K (2011) Like a fish out of water: reconsidering disaster recovery and the role of place and social capital in community disaster resilience. Am J Community Psychol 48(3–4):395–411. https://doi.org/10.1007/s10464-011-9427-0
15. Edwards D (2014) Flooding events in Maidstone: report of director of environment and shared services. Maidstone Borough Council Cabinet, Maidstone, UK. https://services.maidstone.gov.uk/meetings/documents/g2118/Public%20reports%20pack%2008th-Oct-2014%2018.30%20Cabinet.pdf?T=10
16. Environment Agency (2014) Living on the edge. Environment Agency, Bristol, UK. https://www.gov.uk/government/publications/riverside-ownership-rights-and-responsibilities
17. Environment Agency (2015) Risk of flooding from rivers and sea: flood map. http://watermaps.environment-agency.gov.uk/wiyby/wiyby.aspx?topic=floodmap&scale=7&ep=map&layerGroups=default&lang=_e&y=181500&x=531500#x=531502&y=181583&scale=6
18. Farazmand A (2009) Hurricane Katrina, the crisis of leadership, and chaos management: Time for trying the 'surprise management theory in action'.Public Organ Rev 9(4):399–412
19. Fereday J, Muir-Cochrane E (2006) Demonstrating rigor using thematic analysis: a hybrid approach of inductive and deductive coding and theme development. Int J Qual Methods 5(1):80–92. https://doi.org/10.1177/160940690600500107

20. Gale NK, Heath G, Cameron E, Rashid S, Redwood S (2013) Using the framework method for the analysis of qualitative data in multi-disciplinary health research. BMC Med Res Methodol 13(1):117. https://doi.org/10.1186/1471-2288-13-117
21. Galea S, Nandi A, Vlahov D (2005) The epidemiology of post-traumatic stress disorder after disasters. Epidemiol Rev 27(1):78–91. https://doi.org/10.1093/epirev/mxi003
22. Gordon KH, Bresin K, Dombeck J, Routledge C, Wonderlich JA (2011) The impact of the 2009 Red River Flood on interpersonal risk factors for suicide. Crisis J Crisis Interv Suicide 32(1):52–55. https://doi.org/10.1027/0227-5910/a000051
23. Grimm A, Hulse L, Preiss M, Schmidt S (2014) Behavioural, emotional, and cognitive responses in European disasters: results of survivor interviews. Disasters 38(1):62–83. https://doi.org/10.1111/disa.12034
24. Holgersen KH, Boe HJ, Holen A (2010) Long-term perspectives on posttraumatic growth in disaster survivors. J Trauma Stress 23(3):413–416. https://doi.org/10.1002/jts.20530
25. Huang P, Tan H, Liu A, Feng S, Chen M (2010) Prediction of posttraumatic stress disorder among adults in flood district. BMC Public Health 10:207. https://doi.org/10.1186/1471-2458-10-207
26. Lamond J, Joseph RD, Proverbs DG (2015) An exploration of factors affecting the long term psychological impact and deterioration of mental health in flooded households. Environ Res 140:325–334. https://doi.org/10.1016/j.envres.2015.04.008
27. Lane K, Charles-Guzman K, Wheeler K, Abid Z, Graber N, Matte T (2013) Health effects of coastal storms and flooding in urban areas: a review and vulnerability assessment. J Environ Public Health 2013:13. https://doi.org/10.1155/2013/913064
28. Lemyre L, Clément M, Corneil W, Craig L, Boutette P, Tyshenko M et al (2005) A psychosocial risk assessment and management framework to enhance response to CBRN terrorism threats and attacks. Biosecur Bioterror 3(4):316–330. https://doi.org/10.1089/bsp.2005.3.316
29. Lowe SR, Sampson L, Gruebner O, Galea S (2015) Psychological resilience after hurricane sandy: the influence of individual- and community-level factors on mental health after a large-scale natural disaster. PLoS One 10(5):e0125761. https://doi.org/10.1371/journal.pone.0125761
30. Mason V, Andrews H, Upton D (2010) The psychological impact of exposure to floods. Psychol Health Med 15(1):61–73. https://doi.org/10.1080/13548500903483478
31. Murray V, Caldin H, Amlôt R, Stanke C, Lock S, Rowlatt H, Williams R (2011) The effects of flooding on mental health. Health Protection Agency, London. https://webarchive.nationalarchives.gov.uk/ukgwa/20140722082122/http://www.hpa.org.uk/webc/HPAwebFile/HPAweb_C/1317131767423
32. Neria Y, Nandi A, Galea S (2008) Post-traumatic stress disorder following disasters: a systematic review. Psychol Med 38(4):467–480. https://doi.org/10.1017/s0033291707001353
33. Norris FH, Baker CK, Murphy AD, Kaniasty K (2005) Social support mobilization and deterioration after Mexico's 1999 flood: effects of context, gender, and time. Am J Community Psychol 36(1–2):15–28. https://doi.org/10.1007/s10464-005-6230-9
34. Norris FH, Murrell SA (1988) Prior experience as a moderator of disaster impact on anxiety symptoms in older adults. Am J Community Psychol 16(5):665–683. https://doi.org/10.1007/BF00930020
35. Norris FH, Stevens SP, Pfefferbaum B, Wyche KF, Pfefferbaum RL (2008) Community resilience as a metaphor, theory, set of capacities, and strategy for disaster readiness. Am J Community Psychol 41(1–2):127–150. https://doi.org/10.1007/s10464-007-9156-6
36. Paranjothy S, Gallacher J, Amlôt R, Rubin GJ, Page L, Baxter T et al (2011) Psychosocial impact of the summer 2007 floods in England. BMC Public Health 11:145. https://doi.org/10.1186/1471-2458-11-145
37. Patel SS, Rogers MB, Amlôt R, Rubin GJ (2017) What do we mean by 'community resilience'? A systematic literature review of how it is defined in the literature. PLoS Currents 9
38. Patel SS, Webster RK, Greenberg N, Weston D, Brooks SK (2020) Research fatigue in COVID-19 pandemic and post-disaster research: causes, consequences and recommendations. Disaster Prev Manag Int J 29(4):445–455. https://doi.org/10.1108/DPM-05-2020-0164

39. Paton D, Johnston D (2001) Disasters and communities: vulnerability, resilience and preparedness. Disaster Prev Manag Int J 10(4):270–277. https://doi.org/10.1108/EUM0000000005930
40. Paton D, Millar M, Johnston D (2001) Community resilience to volcanic hazard consequences. Nat Hazards 24(2):157–169. https://doi.org/10.1023/A:1011882106373
41. Pfefferbaum B, Pfefferbaum RL, Van Horn RL (2015) Community resilience interventions: participatory, assessment-based, action-oriented processes. Am Behav Sci 59(2):238–253. https://doi.org/10.1177/0002764214550298
42. Phifer JF, Norris FH (1989) Psychological symptoms in older adults following natural disaster: nature, timing, duration, and course. J Gerontol 44(6):S207–217. https://doi.org/10.1093/geronj/44.6.S207
43. QSR International (2012) NVivo qualitative data analysis Software, Version 10: QSR International Pty Ltd.
44. Ritchie J, Lewis J, Nicholls CM, Ormston R (2013) Qualitative research practice: a guide for social science students and researchers, 2 ed. SAGE Publications Ltd., London, UK. https://us.sagepub.com/en-us/nam/qualitative-research-practice/book237434
45. Shinoda T (2013) DPJ's political leadership in response to the Fukushima nuclear accident. Jpn J Polit Sci 14(2):243–259
46. Slingo J, Belcher S, Scaide A, McCarthy M, Saulter A, McBeath K et al (2014) The recent storms and floods in the UK. Met Office, Exeter, UK. http://www.metoffice.gov.uk/media/pdf/n/i/Recent_Storms_Briefing_Final_07023.pdf
47. Stanke C, Murray V, Amlôt R, Nurse J, Williams R (2012) The effects of flooding on mental health: outcomes and recommendations from a review of the literature. PLOS Currents Disasters, 4. https://doi.org/10.1371/4f9f1fa9c3cae
48. Stimpson JP, Wilson FA, Jeffries SK (2008) Seeking help for disaster services after a flood. Disaster Med Public Health Prep 2(3):139–141. https://doi.org/10.1097/DMP.0b013e318183cfbb
49. Tedeschi RG, Calhoun LG (1996) The posttraumatic growth inventory: measuring the positive legacy of trauma. J Trauma Stress 9(3):455–471. https://doi.org/10.1007/bf02103658
50. Tedeschi RG, Calhoun LG (2004) Target article: "Posttraumatic growth: Conceptual foundations and empirical evidence." Psychol Inq 15(1):1–18. https://doi.org/10.1207/s15327965pli1501_01
51. Tedeschi RG, Park CL, Calhoun LG (1998) Posttraumatic growth: positive changes in the aftermath of crisis. Routledge, Mahwah, New Jersey. https://www.routledge.com/Posttraumatic-Growth-Positive-Changes-in-the-Aftermath-of-Crisis/Tedeschi-Park-Calhoun/p/book/9781138002654
52. Thieken AH, Kreibich H, MüLler M, Merz B (2007) Coping with floods: preparedness, response and recovery of flood-affected residents in Germany in 2002. Hydrol Sci J 52(5):1016–1037. https://doi.org/10.1623/hysj.52.5.1016
53. Twigg J (2007) Characteristics of a disaster-resilient community: a guidance note. DFID Disaster Risk Reduction NGO Interagency Coordination Group, London, UK. http://www.preventionweb.net/files/2310_Characteristicsdisasterhighres.pdf
54. Walker G, Burningham K (2011) Flood risk, vulnerability and environmental justice: evidence and evaluation of inequality in a UK context. Crit Soc Policy 31(2):216–240. https://doi.org/10.1177/0261018310396149
55. Werritty A, Houston D, Ball T, Tavendale A, Black A (2007) Exploring the social impacts of flood risk and flooding in Scotland. Scottish Executive, Edinburgh. http://www.gov.scot/Publications/2007/04/02121350/0
56. Wheater HS (2006) Flood hazard and management: a UK perspective. Philos Trans R Soc Math Phys Eng Sci 364(1845):2135–2145. https://doi.org/10.1098/rsta.2006.1817
57. Wilby RL, Beven KJ, Reynard NS (2008) Climate change and fluvial flood risk in the UK: more of the same? Hydrol Process 22(14):2511–2523. https://doi.org/10.1002/hyp.6847

Chapter 3
What Do the Residents of Flood-Risk Areas in the UK Understand About Community Resilience? A Cross-Sectional Study

Abstract This Chapter shares the findings of a 2016 cross-sectional online survey carried out among participants who live in high-risk areas for flooding in the United Kingdom (N = 1,025). According to an exploratory factor analysis, participants interpreted elements of community resilience as three attributes of a community: Strong Fellowship, Trust in the System, and Memories of Disasters. The participants understood a community's psychological, social, and economic impacts caused by severe flooding as adverse outcomes with two impact levels (personal and community). Participants with a strong fellowship in the community or fewer memories of disasters perceived that a potential flood would have a personal impact ($p < 0.05$). In contrast, participants with a weak fellowship in the community, more memories of disasters, or a lack of trust in the system and leadership thought a likely flood would have a community impact ($p < 0.001$). Lastly, participants who felt that they belonged to their community or vaguely knew the location of the local disaster support center in their community thought a severe flood would have no community impact on them ($p < 0.001$). The findings did not suggest the existence of a hierarchy with priorities among community resilience elements.

Keywords Community resilience · Flooding · Risk perception · At-risk population · Public health · Global health · Cross sectional study · Online survey

3.1 Background

In Chap. 2, qualitative evidence was collected on the concept and application of community resilience from people in flood-affected communities in the United Kingdom (UK). The Chap. 2 findings from the focus group discussions supported the taxonomy of the elements of community resilience presented in Patel et al. [38]. The reported connections between the NRR security outcomes (a community's psychological, social, and economic impacts caused by severe flooding) [7] and the elements of community resilience were anecdotally mentioned by a majority of the focus group participants. Based on personal experience with flooding, participants confirmed

S. S. Patel, *Community Resilience When Disaster Strikes*, Advanced Sciences and Technologies for Security Applications,
https://doi.org/10.1007/978-3-031-07992-4_3

that if the activities related to community resilience were applied in their community before the flood, then they would have felt less impact or none of the harmful security and health outcomes, as described by the NRR security outcomes. Replicating these findings in a quantitative survey would further explore and statistically provide possible types of relationships between the elements and the NRR security outcomes. The statistical tests could even highlight the way in which the elements have a positive or negative impact on the NRR security outcomes.

3.1.1 Building upon the Qualitative Research to Further Explore the Relationships Discussed

The details learned from the focus group discussions have established a better foundation for the elements derived from the definitions of community resilience. Further understanding of this foundation can help communities prepare against flooding and reduce the likely impacts efficiently. Each element found by Patel et al. [38] was found necessary by the focus group participants in Chap. 2. An examination in a quantitative way would test and measure whether these elements, seen in the qualitative results, can minimize the overall flood impact in the communities at large after the flood.

The qualitative study did provide deep insights into recent experiences of flooding. The flood experience empowered many focus group participants to mitigate their own future flood risks in many ways. However, it is unclear if people who have not been flooded would have a similar and clear understanding of how resilient or even prepared their community is against such an event. For example, comparing participants living in the areas at a high risk of earthquakes in Romania with participants in low-risk regions of France, Gruev-Vintila and Rouquette [22] found that participants who had higher recognition and practical use of risk information for earthquakes were those who had experienced an earthquake and who were highly active in mitigating earthquake risk. Similarly, Coulston and Deeny [13] found that participants in the UK who experienced a severe flood were better prepared for the next one. But how can practitioners approach the general public living in high flood-risk areas, especially those with no prior disaster experience or even a perception of possible risk? Could the general public interpret the elements of community resilience in the same way as the focus group participants? Answering these questions can help progress and apply the concept of community resilience to flood-risk communities.

3.1.2 Aims of This Chapter

This Chapter aimed to identify how the general public living in high flood-risk areas understands the elements of community resilience and to examine the likely associations between the elements of community resilience and NRR security outcomes if a severe flood occurs in the community.

3.2 Methods

From February 12th, 2016, until March 9th, 2016, a cross-sectional survey was carried out online via TNS Global and targeted people living in high-risk areas for flooding in the UK. The questions relating to the elements of community resilience were based on the results of previous work [38, Chap. 2], while those pertaining to NRR security outcomes were created from the descriptions given by Cabinet [7]. To test the association between elements of community resilience and NRR security outcomes, a portion of the questions asked the participants to answer about their community now and as if they were in a fictional but similar community that has been severely flooded.

3.2.1 Selection Criteria

Participants were selected if they: (1) were older than 16 years old; (2) had access to the internet; (3) had been previously and actively asked, using a double opt-in procedure via TNS Global, to send them online survey notifications; (4) lived in postcodes marked as high flood-risk areas by the UK Environment Agency (EA) [18]; and (5) responded to the screening question 'is your property currently flooded and/or has your property been flooded in the last 6 months?' with an answer of 'no.'

3.2.2 Public Database of Postcodes at Risk of Flooding from Rivers and Seas

Using a public database published by the EA [18] in April 2015, postcodes marked as high-risk areas for flooding from rivers and seas were collected. In total, 41,424 postcodes identified as high flood-risk postcodes in Cumbria, Lancashire, and North Yorkshire were excluded because of the recent floods that occurred in the region during the time of the survey [3].

The calculated risks used in this public database were determined through a series of modeling on the rivers and seas in the UK by the EA teams and included the area's specific flood defenses. These models aimed to" determine how much water would flood the land for a range of events and where it would go" [17, p. 3). High flood-risk regions were identified as having a probability of flooding each year of greater than 1 in 30 (3.3%) [17].

3.2.3 Development and Rationale of Survey Structure and Questions

The survey's overall structure was developed thematically, and questions were grouped into five sections: (a) participants' understanding of 'community'; (b) participants' perceived flood risk and preparedness; (c) items related to the elements of community resilience; (d) items related to the NRR security outcomes using a fictional flooding scenario; and (e) demographics. The learnings from Chap. 2 were also turned into statements for this survey.

For the first section, questions related to the definition of 'community' were included to capture the participants' sense of community. Following this question, a description of the community was provided to the participants to use within the remainder of the survey questions. The broad definition of a community is given as follows: "your local neighborhood comprising your local council leaders, neighbors, and other residents belonging to your parish or borough council."

The second section aimed to learn about the participants' views of their flood risk and preparedness. This section asked the participants about their feeling of belonging to their community, their perception of their flood risk, and their status with flood insurance. Questions on emergency preparedness specific to the UK, taken from a longitudinal study by Page et al. [37], were also included. Page et al. [37] asked the participants what emergency supplies were gathered in their homes and their knowledge of emergency plans.

In the third section, statements related to the elements of community resilience were included. Participants were asked to state whether they strongly disagreed, disagreed, neither agreed nor disagreed, agreed, or strongly agreed with statements identified and generated from quotes gathered from the focus groups. Table 3.1 shows these statements for each element. The actual wording for some statements was adapted from previously used surveys on social capital and support and the assessment given within the Communities Advancing Resilience Toolkit (CART) [26, 31, 39].

As shown in Table 3.1, 29 statements were used for elements of community resilience, and there was at least one question designed for each component. The goal of the statements was to capture the participant's belief in the individual element. For some, direct questions were designed to capture the essence of the element. The unequal number of questions for each was also due to the amount of detail provided by the focus group participants about how certain elements shaped their flooding experience. For example, each focus group provided several diverse specifics of how

Table 3.1 Survey statements developed for each element of community resilience

Element (# of statements)	Statements
Local knowledge (4)	• If it floods, people in my community would have enough local knowledge to cope in the first few days • If it floods, people in my community would have enough local knowledge to cope over the next few months • My community maintains good knowledge regarding potential emergencies that we may face • People in my community share lessons on how to cope with flooding
Communication (5)	• People in my community discuss issues so that we can improve the community • My community keeps people informed (via television, radio, newspaper, internet, phone, neighbours, etc.) about issues that are relevant to them • If a disaster occurs, my community would provide information about what to do • My community has mechanisms for routinely providing accurate information to residents about local issues • People in my community trust the local news media to deliver accurate information
Community networks and relationships (6)	• People in my community feel like they belong to the community • People in my community help each other • People in my community work together to improve the community • People in my community trust one another • People in my community trust public officials • If something bad happened to only a few homes in the community, the rest of the people in the community would help those affected
Governance and leadership (4)	• My community leaders would be effective in a disaster • Help from officials is available to my community if an emergency happens • If a disaster happens, I would expect my community to be left on its own to deal with it • If an emergency happens in my community, I know who will be in charge
Preparedness (1)	• My community actively prepares for future disasters
Mental outlook (4)	• People in my community have hope about the future • I am optimistic in what the future has in store for my community • My community is still traumatised by previous incidents or disasters • My community can cope with a future disaster
Health (3)	• My community is aware of those who would need help in time of an emergency due to an existing health condition • The health services available in my community are high quality • The health services available in my community would be quick to react to any disaster
Economic investment (1)	• My community has the financial resources it needs to take care of problems if affected by a disaster
Resources (1)	• Any aid and help would be distributed justly in my community if a disaster happens

working together and helping each other in the community reduced the participants' flood impact.

The participants were asked to read the following hypothetical flooding scenario in the fourth section and were given speculative flooding images alongside:

> In the fictional scenario described below, we will ask questions about how your community would cope in recovering from a severe flood. Just to reassure you, we do not have any reason to think that a flood is going to happen where you live. We are running this survey to test out theories on behalf of an academic team at King's College London.
>
> Pretend you live in a fictional town that mirrors the characteristics of your current residence, neighbors, and town. Your home and your street were affected by a severe flood with at least knee-high water caused by winter storms. Incoming and outgoing access through the street via cars is not possible, and basic utilities, such as indoor plumbing and electricity, are unreliable. There is no indication on when the floodwater will be going away, as rain continues to fall.

After reading this scenario, the participants were presented with statements related to the NRR security outcomes (Table 3.2). Similar to the third section, participants were asked to state whether they strongly disagreed, disagreed, neither agreed nor disagreed, agreed, or strongly agreed with each statement. The twelve statements on social disruption, psychological impact, and economic harm were developed using the descriptions given by UK Cabinet Office [7] and expanding on these descriptions with details received from the focus group discussions on the flood experience. For example, the focus group participants said they felt high levels of anxiety and depression after the flood when asked about their psychological impact. Statements

Table 3.2 Survey statements for each NRR security outcome

NRR security outcomes (# of statements)	Statements
Psychological impact (4)	• Because of the flood, I would probably feel nervous, anxious or on edge • Because of the flood, I would probably not be able to stop or control worrying • Because of the flood, I would probably have little interest or pleasure in doing things • Because of the flood, I would probably feel down, depressed, or hopeless
Social disruption (5)	• Because of the flood, I would probably stop some social community activities • Because of the flood, I would probably leave or want to leave this community for good • The flood would probably change my community's way of life • The flood would probably make my community a worse place to live in short term • The flood would probably make my community a worse place to live in long term
Economic harm (3)	• Because of the flood, I would probably have financial problems or worries • Because of the flood, I would expect that my house, property, or rent value would go down • Because of the flood, my household income would be at jeopardy

for psychological impact were taken from the validated Patient Health Questionnaire for Depression and Anxiety (PHQ-4) [29, 32]. Statements from the focus group discussions and items developed by Kaniasty [26] were also used to create statements about economic harm and social disruption.

Section five asked for the participants' gender, age, ethnic group, household income, length in the community, previous flood experience, education level, and knowledge of the local community emergency center. In total, there were 55 questions to fit into a time limit of 20 min.

3.2.4 *Piloting the Survey Questions*

After creating the survey questions, a pilot test was carried out with people who participated in the focus groups from Chap. 2. This document tried to mimic how the questions would be styled online. Nine participants who had indicated a desire to help with further research related to flooding were asked for feedback; six participants replied with input and/or a completed survey questionnaire. Although some participants had initial issues, all six participants provided input on the following questions: (a) estimated time it would take them to complete the survey online, (b) difficulty of answering the questions, and (c) questions to add, remove, or revise. Appropriate changes and edits to the survey were made based on this feedback.

3.2.5 *Survey Procedure*

TNS Global created the online survey on their platform upon receiving the final copy of the survey questions. The survey's information sheet included the nature and purpose of the study and stated that the survey was being conducted on behalf of researchers from King's College London (KCL). The information sheet made it clear to potential participants that there was no obligation to participate in the survey. If they choose to participate, they could withdraw at any time by closing the online window. Participants were also informed that the survey followed the requirements outlined by the UK Data Protection Act 1998, the Market Research Society, and KCL Research Ethics. Contact details were given on the information sheet page if potential participants wanted further information or had questions before proceeding. If potential participants still wished to participate in the survey, they clicked "yes". They then reached the survey's screener question of 'is your property currently flooded and/or has your property been flooded in the last 6 months?' If "yes" was selected to this screener question, the participants were thanked for their time and were directed to helpful online resources related to flooding on the EA and Public Health England websites. If "no" was answered to the screening question, the participants confirmed their postcode and were directed to part 1 of the online survey if the postcode was marked as high-risk areas for flooding from rivers and seas in a previously established database.

3.2.6 Number of Participants and Response Rate

The goal was to have a sufficiently large sample to have a high degree of confidence in the results. To find the number of survey participants needed to achieve this, the methods and calculations by the UK National Audit Offices were followed [42, pp. 8–9]. The assumption was that the survey questions would result in a normal distribution of outcomes. Based on the most conservative estimate, assumed dichotomized outcomes from the survey (agree or disagree) at 50% of the UK population. This assumption calculated that having 1,067 participants would provide a 95% confidence interval of plus or minus 3 percent for each result estimate.

However, only 1,025 people who lived in the desired postcodes and met the selection criteria were recruited and responded. The sample size of this survey was short of the goal by 42 people, but the precision of this sample size was still good, providing a 95% confidence interval of plus or minus of 3.1%.

3.2.7 Analysis

Responses from the survey were analyzed with IBM SPSS Statistics and Microsoft Excel. One question had to be dropped from the analysis due to a technical error in its phrasing. Two independent maximum-likelihood exploratory factor analyses using varimax rotation with Kaiser normalization were carried out on the statements related to elements of community resilience and those related to NRR security outcomes. The maximum-likelihood exploratory factor analysis was used over principal component analysis because the magnitude of the loadings found with the maximum-likelihood extraction has a lower tendency to inflate the loading values for each item within each factor [20]. This tendency meant that the items loaded onto the factor were more likely closer to grouping a common factor than those found using principal component analysis.

Because the objective was to explore the relationships between all of the elements of community resilience and NRR security outcomes, all statements (items) were designated to a factor according to its highest loading value, which went against the standard rule of thumb. The standard rule of thumb for having "good" factors suggests that each statement should have a loading value of at least 0.4 and have a difference in loading values of greater than 0.2 between factors. Conventionally, the statements that do not meet these criteria should be dropped from a factor. However, all the statements in the analysis were chosen and kept because this type of approach would best explore and represent the overlaps found within the elements and NRR security outcomes. For example, if a statement had a loading value of 0.329 for Factor

1 and 0.232 for Factor 2, then this statement was designated to Factor 1. This also meant that not all statements loaded uniquely on one factor, which meant that the results would not be useful for creating a scale or questionnaire unless the standard rule of thumb was applied. However, keeping all the statements and using the highest loading value would best represent the elements of community resilience found in the focus groups and explore the organizing principles among the elements and NRR security outcomes as to how the high flood-risk general public saw them.

The scree plot of the exploratory factor analysis and the factor's eigenvalue were considered when determining the number of extracted factors. The factors from the loaded statements were labelled and identified as the respective common theme found in the grouping of the statements. Factor scores using the regression method were estimated for each individual factor based on a weighted sum of each statement in the factor. These scores reflected the contribution of the overlapping statements, but they were weighted in proportion to their correlation to the factor.

Linear regressions were performed between the factor scores found for the elements of community resilience and the factor scores found for the NRR security outcomes. The association between the factor scores found for the NRR security outcomes and the demographic and flood-related variables was also tested. All regressions were conducted with and without controlling for sex, age, region, education, ethnicity, and household income. The values of the unstandardized B coefficient (and its confidence interval) and the standardized Beta coefficient (and its *p*-value) were calculated.

3.2.8 *Ethics*

KCL Research Ethics Committee approved the cross-sectional survey used in this chapter (refs: HR-15/16–1969 and RESCMR-15/16–1969).

3.3 Results

Figure 3.1 shows the flow chart of the recruitment of the participants for the survey. Out of 1,025 participants, 52% were female, and 48% were male. Most participants were aged between 16 and 45 (57%), had household incomes of below £49,000 (65%), and identified themselves as ethnically white (90%). Most had lived in their local community or town for longer than seven years (65%) and said where they currently lived was not at risk of flooding (71%). Based on their postal regions found by their postcodes, more participants lived in the south-east and south-west regions of the UK than the other regions (21% and 15%, respectively). The participants' demographic and flood-related background variables are presented in Table 3.3.

When asked about items already gathered in their home in case of an emergency, 662 participants said they had a torch (65%), 571 participants had candles

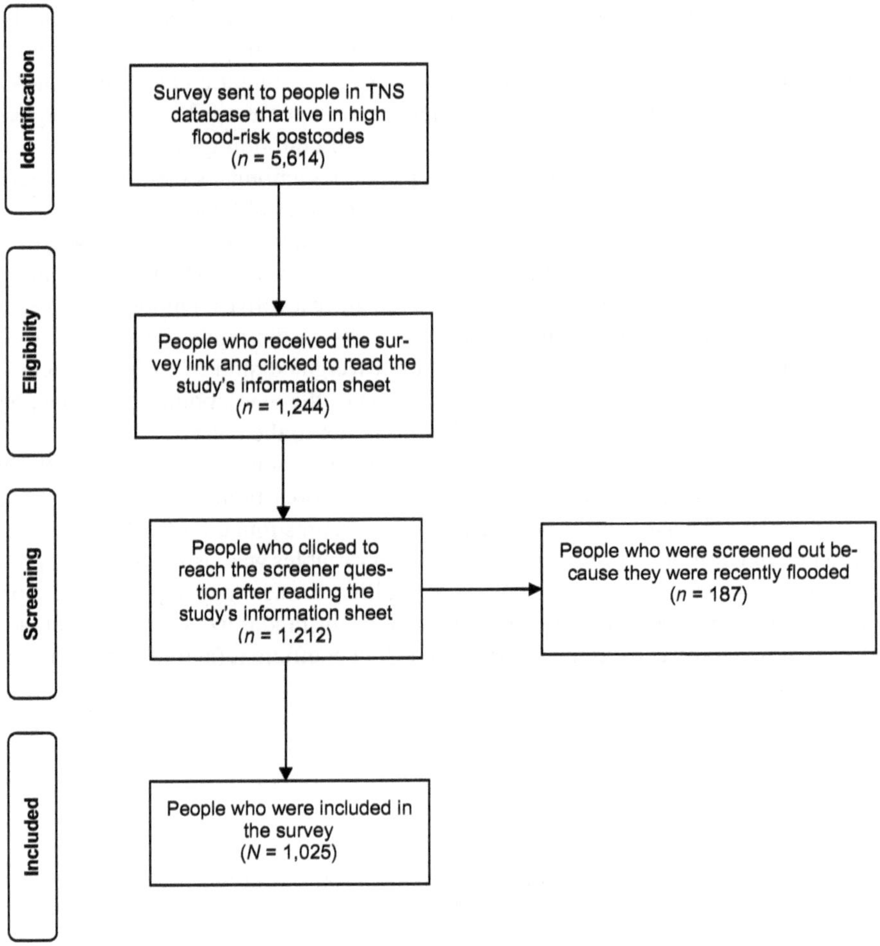

Fig. 3.1 Flow chart of the participant recruitment for the cross-sectional survey

and matches (56%), and 547 participants had first aid kits (53%) in their homes. Two hundred six participants (20%) said they had none of the listed items in their homes. When asked if their property had ever been directly affected by floods, 89% of the participants said 'no,' and 52% of the participants were not sure or did not have such insurance when asked about flood insurance. Among those who said they had flood insurance, 67% of the participants had both building and contents insurance. Additionally, only a small minority of participants said they knew the location of their community's local disaster support center (16%), but 45% said they could guess where this location could be.

Concerning the participant's perception of 'community,' the strongest agreement from the choices offered was "immediate neighbors" (i.e., people who "live next door or on your road"), with which 813 participants (79%) agreed or strongly agreed with.

Table 3.3 Demographic and flood-related background variables

Variables	Levels	Number (%)
Sex	Male	492 (48)
	Female	553 (52)
Age	16 to 30	251 (25)
	31 to 50	425 (41)
	51+	349 (34)
Region	Greater London	124 (12)
	Scotland and North England	194 (19)
	Midlands	225 (22)
	East Anglia	113 (11)
	South East	211 (21)
	South West	158 (15)
Education	Secondary education and lower	536 (52)
	Higher education and above	489 (48)
Ethnicity	White	926 (90)
	Other	99 (10)
Household income	£0 to 25,000	354 (35)
	£25,000 to 49,000	313 (30)
	£49,000+	228 (22)
	Rather not say	130 (13)
Flood insurance	No	326 (32)
	Yes	492 (48)
	Not sure	207 (20)
Length in community	0 to 4 years	281 (27)
	5 to 10 years	228 (22)
	11 to 20 years	262 (26)
	21 + years	254 (25)
Previous flooding	No	913 (89)
	Yes	112 (11)
Perceived flood risk	No	730 (71)
	Yes	200 (20)
	Don't know	95 (9)
Perceived belonging to community	Not at all or not very strongly	500 (49)
	Fairly or very strongly	525 (51)
Knowing location of local disaster support centre	No, don't' know	406 (40)
	Yes	159 (16)
	No, but have guess	460 (45)

A total of 765 participants (75%) included their "local friends," and 684 participants (67%) had their "wider neighbors (e.g., people in the same town/district)" in their interpretation of community. The most substantial disagreement among the choices was "people that I know online," which 522 participants (51%) disagreed or strongly disagreed with. Other strong disagreements were found with the statements related to their "religious or faith-based group" and their "non-local friends," which 447 participants (44%) and 413 participants (40%) respectively disagreed or strongly disagreed with.

Overall, 577 participants (56%) agreed that people in their community help each other, but 492 participants (48%) disagreed that people in their community share lessons on coping with flooding. If an area is flooded, only 517 participants (50%) agreed that their community would have enough local knowledge to manage in the first few days. Whether their community leaders would be effective in a disaster, 544 participants (53%) neither agreed nor disagreed with that statement. Furthermore, 434 participants (42%) expected their community not to be left on its own to deal with a disaster if it happens, although 530 participants (52%) said they did not know who would be in charge of an emergency if happening in their community. Nevertheless, 584 participants (57%) believed that their community members would help if something bad happened to only a few homes in their community. Additionally, 572 participants (56%) felt the available health services in their community were high quality, and 630 participants (61%) thought that their community was no longer traumatized by previous incidents or disasters.

When the participants were asked to consider a potential flood and its likely psychological impact, over half of the participants agreed or strongly agreed with the PHQ-4 statements, except for one statement related to depression, for which only 46% were in agreement. A total of 765 participants (75%) expected that their house, property, or rent value would decrease in a likely flood, and 647 participants (63%) said they would most likely have financial problems or worries because of the flood. In a likely flood scenario, 513 participants (50%) would probably stop some social community activities after the flood, and 646 participants (63%) agreed that such a flood would change their community's way of life. A large proportion of the participants (64%) felt their community would be a worse place to live in the short term after a likely flood. A mixed result was found when asked whether the participant's community would be a worse place to live in the long term after a likely flood (35% in agreement, 31% in disagreement, 34% neither agree nor disagree).

3.3.1 Exploratory Factor Analysis of Statements Related to the Elements of Community Resilience

As shown in the scree plot in Fig. 3.2, the analysis initially found four factors. Each factor had an eigenvalue greater than one, but the slope of the curve of the scree plot appeared to be less steep and plateaued on factor 4. The first three factors explained

Fig. 3.2 Scree plot of the factors found among statements related to elements of community resilience

a total of 54% of the variance for the entire set of statements related to the elements of community resilience. In contrast, the fourth factor only explained 3% of the total variance seen. Because of these findings, the statements seemed to fit better into three factors instead of 4. Table 3.4 shows the calculated loading values (sorted by largest value) for each statement based on limiting the choices to three factors. Eleven statements were placed in the first factor using the largest loading value between the factors, which explained 43% of the total variance seen and by far the most. Twelve statements were placed in the second, which explained 6% of the total variance, and six statements were placed in the third, explaining 5% of the total variance. Table 3.5 shows the breakdown of the elements of community resilience with respect to each factor. Some of the statements overlapped between factors, which are italicized in Table 3.5. Factor 2 was less well determined than factors 1 and 3 due to the overlapping statements. The factors were seen as Strong Fellowship (first factor), Trust in the System (second factor), and Memories of Disasters (third factor).

Table 3.4 Results of exploratory factor analysis on the elements of community resilience, sorted by largest loading value

Rotated factor matrix[a]			
Statements	Factor		
	1	2	3
People in my community help each other	**0.802**	0.247	
People in my community feel like they belong to the community	**0.721**	0.302	
People in my community trust one another	**0.713**	0.274	
People in my community work together to improve the community	**0.682**	0.354	
If something bad happened to only a few homes in the community, the rest of the people in the community would help those affected	**0.651**	0.298	
People in my community have hope about the future	***0.547***	*0.464*	
People in my community discuss issues so that we can improve the community	***0.535***	*0.324*	*0.365*
I am optimistic in what the future has in store for my community	***0.514***	*0.463*	
My community is aware of those who would need help in time of an emergency due to an existing health condition	***0.500***	*0.399*	*0.254*
If it floods, people in my community would have enough local knowledge to cope over the next few months	***0.461***	*0.453*	*0.221*
If it floods, people in my community would have enough local knowledge to cope in the first few days	***0.446***	*0.411*	
The health services available in my community would be quick to react to any disaster		**0.651**	
My community leaders would be effective in a disaster	0.349	**0.596**	0.223
Help from officials is available to my community if an emergency happens	0.280	**0.594**	
The health services available in my community are high quality	0.214	**0.582**	
People in my community trust public officials	0.313	**0.542**	
People in my community trust the local news media to deliver accurate information	0.224	**0.541**	
My community has the financial resources it needs to take care of problems if affected by a disaster	0.287	**0.526**	0.282
Any aid and help would be distributed justly in my community if a disaster happens	*0.418*	***0.525***	
If a disaster occurs, my community would provide information about what to do	*0.446*	***0.520***	
My community can cope with a future disaster	*0.413*	***0.516***	
*My community keeps people informed (*via *television, radio, newspaper, internet, phone, neighbours, etc.) about issues that are relevant to them*	*0.433*	***0.482***	
My community has mechanisms for routinely providing accurate information to residents about local issues	*0.450*	***0.463***	*0.283*
People in my community share lessons on how to cope with flooding	0.318	0.251	**0.675**

(continued)

Table 3.4 (continued)

Rotated factor matrix[a]			
Statements	Factor		
	1	2	3
My community is still traumatised by previous incidents or disasters			**0.650**
My community actively prepares for future disasters	0.294	0.372	**0.612**
If an emergency happens in my community, I know who will be in charge	*0.294*	*0.385*	***0.490***
My community maintains good knowledge regarding potential emergencies that we may face	*0.438*	*0.406*	***0.446***
If a disaster happens, I would expect my community to be left on its own to deal with it			**0.282**

Note Factor loadings < 0.200 are suppressed; italics- statements that did not load uniquely
Extraction Method: Maximum Likelihood
Rotation Method: Varimax with Kaiser Normalization
[a] Rotation converged in 5 iterations

Table 3.5 Overview of the loaded statements related to the elements of community for each factor

Factor, Cronbach's alpha (α), Eigenvalue	Number of loaded statements	Elements (# of loaded statements)
1. Strong fellowship, $\alpha = 0.920$, Eigenvalue = 12.436	11	Local knowledge (2) Communication (1) Community networks and relationships (5) Mental outlook (2) Health (1)
2. Trust in system, $\alpha = 0.900$, Eigenvalue = 1.861	12	Health (2) Governance and leadership (2) Communication (4) Community networks and relationships (1) Economic (1) Resources (1) Mental outlook (1)
3. Memories of disasters, $\alpha = 0.776$, Eigenvalue = 1.355	6	Local knowledge (2) Governance and leadership (2) Preparedness (1) Mental outlook (1)

3.3.2 Strong Fellowship

The first factor combined statements that asked participants whether they thought their community could handle and cope with flooding. The common thread among the statements for this factor was the sense of Strong Fellowship in the community. Strong Fellowship included statements related to five elements: local knowledge, communication, community networks and relationships, mental outlook, and health.

The element of community networks and relationships had the greatest number of statements, which were also the largest loading values in this factor. Strong fellowship demonstrated good internal consistency: the Cronbach alpha (α) was 0.920, and the eigenvalue was 12.436, with item loads ranging from 0.446 to 0.802. Not all statements loaded uniquely on this factor. The specific statements loaded in Strong Fellowship are provided in Table 3.6.

3.3.3 Trust in the System

Trust in the people in charge and the current system in place was seen as the common thread among the statements loaded onto the second factor. The 'system' in place was inferred as the community's protocols and procedures in place for flooding. The statements loaded onto this factor consisted of six elements: health, government and leadership, communication, community networks and relationships, economic investment, and resources. Trust in the System showed good internal consistency: the Cronbach alpha (α) was 0.900, and the eigenvalue was 1.861, with item loads ranging from 0.463 to 0.651, but again not all statements were loaded uniquely. The specific statements that loaded onto Trust in System are provided in Table 3.7.

3.3.4 Memories of Disasters

The third factor was Memories of Disasters, which encompassed statements that broadly asked the participants about their memories of past disasters and their knowledge on future ones. The statements with greater weight in this factor asked participants if they were still traumatized by past disasters, if their community members share lessons on coping with flooding, and whether their community is actively preparing for future disasters. This factor consisted of four elements: local knowledge, governance and leadership, preparedness, and mental outlook. Naming this factor was less straightforward than the others; however, the consistent theme from the statements stemmed more from asking participants about having a memory of a flood rather than about coping strategies or specific knowledge. This factor showed fair internal consistency: the Cronbach alpha (α) was 0.776, and the eigenvalue was 1.355, with item loads ranging from 0.282 to 0.675, but not all statements loaded uniquely, and one statement was below the loading value of 0.4. The statements that loaded onto this factor are provided in Table 3.8.

Table 3.6 Statements that loaded onto Strong Fellowship, sorted by largest loading value per element

Element	Survey statements (Loading value)	Number of people who… (%)		
		Strongly agree or agree	Neither agree or disagree	Strongly disagree or disagree
Community networks and relationships	People in my community help each other (0.802)	577 (56)	307 (30)	141 (14)
	People in my community feel like they belong to the community (0.721)	444 (43)	414 (41)	167 (16)
	People in my community trust one another (0.713)	446 (44)	393 (38)	186 (18)
	People in my community work together to improve the community (0.682)	485 (47)	371 (36)	169 (17)
	If something bad happened to only a few homes in the community, the rest of the people in the community would help those affected (0.651)	584 (57)	300 (29)	141 (14)
Mental outlook	People in my community have hope about the future (0.547)	431 (42)	473 (46)	121 (12)
	I am optimistic in what the future has in store for my community (0.514)	460 (45)	415 (40)	150 (15)
Communication	People in my community discuss issues so that we can improve the community (0.535)	353 (34)	360 (35)	312 (31)
Health	My community is aware of those who would need help in time of an emergency due to an existing health condition (0.500)	368 (36)	425 (41)	232 (23)
Local knowledge	If it floods, people in my community would have enough local knowledge to cope over the next few months (0.461)	457 (45)	372 (36)	196 (19)
	If it floods, people in my community would have enough local knowledge to cope in the first few days (0.446)	517 (50)	335 (33)	173 (17)

Table 3.7 Statements that loaded onto Trust in System, sorted by largest loading value per element

Element	Survey statements (Loading value)	Number of people who… (%)		
		Strongly agree or agree	Neither agree or disagree	Strongly disagree or disagree
Health	The health services available in my community would be quick to react to any disaster (0.651)	480 (47)	380 (37)	165 (16)
	The health services available in my community are high quality (0.582)	572 (56)	299 (29)	154 (15)
Governance and leadership	My community leaders would be effective in a disaster (0.596)	277 (27)	544 (53)	204 (20)
	Help from officials is available to my community if an emergency happens (0.594)	417 (41)	468 (46)	140 (13)
Community networks and relationships	People in my community trust public officials (0.542)	287 (28)	506 (49)	232 (23)
Communication	People in my community trust the local news media to deliver accurate information (0.541)	468 (46)	413 (40)	144 (14)
	If a disaster occurs, my community would provide information about what to do (0.520)	478 (47)	388 (37)	159 (16)
	My community has mechanisms for routinely providing accurate information to residents about local issues (0.463)	412 (40)	395 (39)	218 (21)
	My community keeps people informed (via television, radio, newspaper, internet, phone, neighbors, etc.) about issues that are relevant to them (0.482)	488 (48)	327 (32)	210 (20)
Economic investment	My community has the financial resources it needs to take care of problems if affected by a disaster (0.526)	247 (24)	475 (46)	303 (30)
Resources	Any aid and help would be distributed justly in my community if a disaster happens (0.525)	440 (43)	446 (44)	139 (13)
Mental outlook	My community can cope with a future disaster (0.516)	406 (40)	464 (45)	155 (15)

Table 3.8 Statements that loaded onto Memories of Disasters, sorted by largest loading value per element

Element	Survey statements (Loading value)	Number of people who... (%)		
		Strongly agree or agree	Neither agree or disagree	Strongly disagree or disagree
Local knowledge	People in my community share lessons on how to cope with flooding (0.675)	200 (20)	333 (32)	492 (48)
	My community maintains good knowledge regarding potential emergencies that we may face (0.446)	342 (33)	463 (45)	220 (22)
Mental outlook	My community is still traumatized by previous incidents or disasters (0.650)	134 (13)	261 (25)	630 (62)
Preparedness	My community actively prepares for future disasters (0.612)	231 (23)	383 (37)	411 (40)
Governance and leadership	If an emergency happens in my community, I know who will be in charge (0.490)	196 (19)	299 (29)	530 (52)
	If a disaster happens, I would expect my community to be left on its own to deal with it (0.282)	265 (26)	326 (32)	434 (42)

3.3.5 Findings Based on Exploratory Factor Analysis of the NRR Security Outcomes

The analysis yielded two factors explaining a total of 53% of the variance for the entire set of statements related to the NRR security outcomes. As shown in the scree plot in Fig. 3.3, both factors had an eigenvalue greater than 1, and the slope of the curve of the scree plot looked to be less steep and plateaued after factor 2. Using only

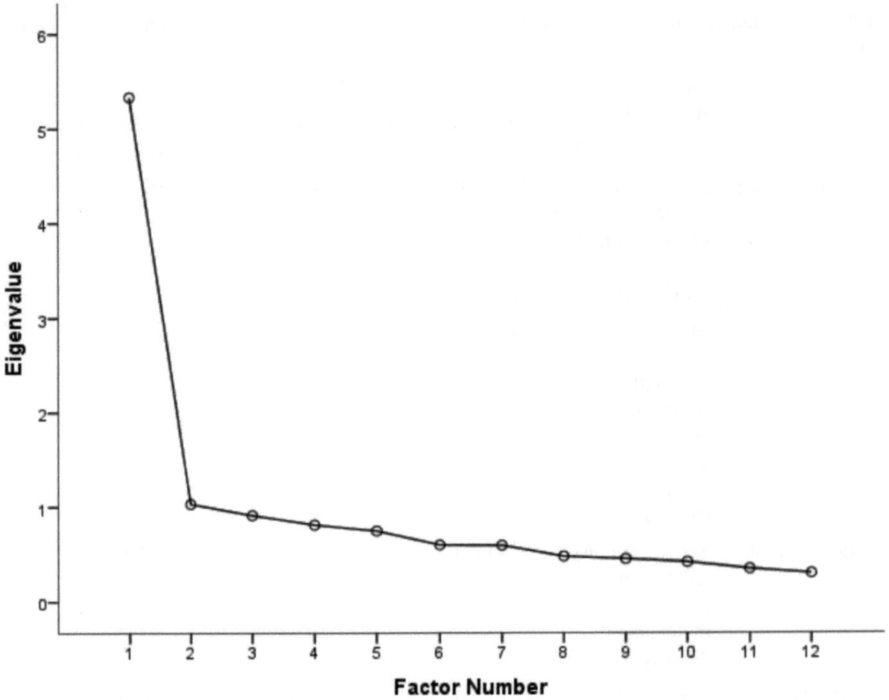

Fig. 3.3 Scree plot of the factors found among NRR security outcomes

one factor could have been justified as well. Still, since this analysis was explorative, two factors were decided because 'over factoring' has been shown by Fabrigar et al. [20] to be more likely to accurately identify factors than under-factoring. Thus, using two factors instead of one would increase the accuracy of the factors seen.

Table 3.9 presents the loadings for each statement, sorted by the largest loading values first and with statements italicized that did not load uniquely. Eight statements, which consisted of four overlapping statements with factor 2, were placed in the first factor and explained 44% of the total variance seen. Explaining 9% of the total variance, the second factor included four statements, which had one statement overlapping with factor 1. Table 3.10 shows the breakdown of the NRR security outcomes with respect to each factor. The first factor was seen as Personal Impact, while the second one was seen as Community Impact.

3.3.6 *Personal Impact*

Personal Impact can be best summarized as the individual negative impacts that participants perceived a potential flood would cause. For example, scoring higher in this outcome factor meant that the participants believed a possible flood would

Table 3.9 Results exploratory factor analysis on NRR security outcomes, sorted by largest loading value

Rotated factor matrix[a]		
Statements	Factor	
	1	2
Because of the flood I would probably feel nervous, anxious or on edge	**0.765**	
Because of the flood I would probably not be able to stop or control worrying	**0.754**	0.319
Because of the flood I would probably feel down, depressed, or hopeless	**0.727**	0.328
Because of the flood I would probably have financial problems or worries	**0.572**	0.282
Because of the flood I would probably have little interest or pleasure in doing things	***0.543***	*0.408*
Because of the flood my household income would be at jeopardy	***0.449***	*0.356*
Because of the flood I would expect that my house, property, or rent value would go down	***0.444***	*0.267*
The flood would probably change my community's way of life	***0.429***	*0.332*
The flood would probably make my community a worse place to live in the long term	0.217	**0.761**
Because of the flood I would probably leave or want to leave this community for good	0.250	**0.635**
The flood would probably make my community a worse place to live in the short term	0.376	**0.506**
Because of the flood I would probably stop some social community activities	*0.289*	***0.349***

Note Factor loadings < 0.200 are suppressed; italics- statements that did not load uniquely
Extraction Method: Maximum Likelihood
Rotation Method: Varimax with Kaiser Normalization
[a] Rotation converged in 3 iterations

Table 3.10 Number of statements related to NRR security outcomes that loaded onto Factors

Factor, Cronbach's alpha (α), Eigenvalue	Number of loaded statements	NRR security outcomes (Number of loaded statements)
1. Personal impact, $\alpha = 0.863$, Eigenvalue = 5.332	8	Psychological impact (4) Economic harm (3) Social disruption (1)
2. Community impact, $\alpha = 0.716$, Eigenvalue = 1.031	4	Social disruption (4)

increase their anxiety levels and feelings of financial insecurity. As seen in Table 3.11, this factor consisted of all the statements associated with NRR security outcomes of psychological impact and economic harm, along with one statement related to social disruption. Personal Impact showed good internal consistency: the Cronbach alpha

Table 3.11 Statements loaded onto Personal Impact, sorted by largest loading value per NRR security outcome

NRR security outcome	Survey statements (Loading value)	Number of people who… (%)		
		Strongly agree or agree	Neither agree or disagree	Strongly disagree or disagree
Psychological impact	Because of the flood I would probably feel nervous, anxious or on edge (0.765)	713 (70)	190 (19)	122 (12)
	Because of the flood I would probably not be able to stop or control worrying (0.754)	555 (54)	255 (25)	215 (21)
	Because of the flood I would probably feel down, depressed, or hopeless (0.727)	614 (60)	232 (23)	179 (17)
	Because of the flood I would probably have little interest or pleasure in doing things (0.543)	467 (46)	293 (29)	265 (26)
Economic harm	Because of the flood I would probably have financial problems or worries (0.572)	647 (63)	212 (21)	166 (16)
	Because of the flood my household income would be at jeopardy (0.449)	466 (45)	260 (25)	299 (29)
	Because of the flood I would expect that my house, property, or rent value would go down (0.444)	765 (75)	186 (18)	74 (7)
Social disruption	The flood would probably change my community's way of life (0.429)	646 (63)	250 (24)	129 (13)

(α) was 0.863, and the eigenvalue was 5.332, with item loads ranging from 0.429 to 0.765, but only half of the statements loaded uniquely on this factor.

3.3.7 Community Impact

The participants perceived that a potential flood would harm their community, a common thread within the second outcome factor, Community Impact. As seen in Table 3.12, this factor consisted of statements asking if participants thought their community would be a worse place to live and if they would want to or probably decide to leave their community after a severe flood. Only 25% of the total statements designed for the NRR Security outcomes were seen in the factor of Community Impact, and these statements were all related to social disruption. Community Impact demonstrated fair internal consistency: the Cronbach alpha (α) was 0.716, and the eigenvalue was 1.031, with item loads ranging from 0.329 to 0.761, but only one statement was not loaded uniquely.

Table 3.12 Statements loaded onto Community Impact, sorted by largest loading value

NRR security outcome	Survey statements (Loading value)	Number of people who… (%)		
		Strongly agree or agree	Neither agree or disagree	Strongly disagree or disagree
Social disruption	The flood would probably make my community a worse place to live in the long term (0.761)	356 (35)	350 (34)	319 (31)
	Because of the flood I would probably leave or want to leave this community for good (0.635)	359 (35)	289 (28)	377 (37)
	The flood would probably make my community a worse place to live in the short term (0.506)	656 (64)	235 (23)	134 (13)
	Because of the flood I would probably stop some social community activities (0.349)	513 (50)	317 (31)	195 (19)

3.3.8 *Linear Regression Analysis*

Tables 3.13, 3.14, and 3.15 provide the findings from the linear regression analyses. The adjusted values in the tables were controlled for sex, age, region, education, ethnicity, and household income.

3.3.9 *Associations with Personal Impact Factor Scores, Adjusted for Controls*

When adjusting for age, gender, region, income, and ethnicity, participants older than 51 years or with a household income greater than £49,000 were less likely to perceive that a flood would hurt them personally. Using the same controls, participants who were female or living in the Midlands region of the UK were more likely to perceive that a flood would have a personal impact. Participants with a stronger belief in fellowship in the community were more likely to perceive that a flood would have a personal impact. In contrast, participants with more memories of disasters were less likely to perceive a flood would have such an impact. No significant association with Personal Impact was found for the following: education, ethnicity, flood insurance, length of time as part of the community, previous flooding, perception of belonging to a community, awareness of the location of local disaster support centers, and Trust in the System.

3.3.10 *Associations with Community Impact Factor Scores, Adjusted for Controls*

When adjusting for age, gender, region, income, and ethnicity, participants who were residing in the south-west area of the UK were less likely to perceive a flood would have a negative community impact. In contrast, ethnic minority participants were more likely to think a flood would have such an adverse community impact. Participants with a stronger belief of fellowship in the community or who had a higher trust in the system were less likely to perceive a flood would have a community impact. However, the participants who had more memories of disasters were more likely to perceive that a flood would adversely impact their community. Additionally, participants who felt that they belonged to their community or had a guess or knew the location of their local disaster support center were less likely to perceive a flood would have a community impact. Non-significant associations with Community Impact were found for the following: education, household income, flood insurance, length of time as part of the community, previous flooding experience, and perceived flood risk.

Table 3.13 Demographic variables and its associations with factor scores related to NRR security outcomes ($N = 1{,}025$)

Variables	Levels	Number (%)	*Personal impact*				*Community impact*			
			Unadjusted B [95% CI]	Unadjusted β	Adjusted[a] B [95% CI]	Adjusted[a] β	Unadjusted B [95% CI]	Unadjusted β	Adjusted[a] B [95% CI]	Adjusted[a] β
Sex	Male	492 (48)	Reference	Reference	Reference	Reference	Reference	Reference	Reference	Reference
	Female	553 (52)	0.392 [0.255, 0.470]	0.203***	0.352 [0.244, 0.461]	0.197***	−0.067 [−0.170, 0.035]	−0.040	−0.051 [−0.154, 0.051]	−0.031
Age	16 to 30	251 (25)	Reference	Reference	Reference	Reference	Reference	Reference	Reference	Reference
	31 to 50	425 (41)	0.004 [−0.135, 0.144]	0.002	0.006 [−0.133, 0.146]	0.004	−0.027 [−0.176, 0.084]	−0.046	0.023 [−0.110, 0.155]	0.013
	51+	349 (34)	−0.180 [−0.324, −0.035]	−0.095*	−0.192 [−0.338, −0.046]	−0.102**	−0.086 [−0.286, −0.015]	−0.151*	−0.067 [−0.205, 0.072]	−0.038
Region	Greater London	124 (12)	Reference	Reference	Reference	Reference	Reference	Reference	Reference	Reference
	Scotland and North England	194 (19)	0.220 [0.019, 0.421]	0.096*	0.167 [−0.038, 0.372]	0.073	−0.166 [−0.353, 0.021]	−0.078	−0.096 [−0.290, 0.099]	−0.045
	Midlands	225 (22)	0.231 [0.036, 0.426]	0.107*	0.200 [0.002, 0.39]	0.093*	−0.067 [−0.250, 0.115]	−0.033	−0.009 [−0.197, 0.179]	−0.004
	East Anglia	113 (11)	−0.084 [−0.311, 0.143]	−0.030	−0.117 [−0.346, 0.113]	−0.041	−0.295 [−0.506, −0.083]]	−0.111**	−0.215 [−0.432, 0.003]	−0.081
	South East	211 (21)	0.121 [−0.076, 0.319]	0.055	0.119 [−0.079, 0.317]	0.054	−0.124 [−0.308, 0.060]	−0.060	−0.067 [−0.255, 0.121]	−0.032

(continued)

Table 3.13 (continued)

Variables	Levels	Number (%)	*Personal impact*				*Community impact*			
			Unadjusted B [95% CI]	Unadjusted β	Adjusted[a] B [95% CI]	Adjusted[a] β	Unadjusted B [95% CI]	Unadjusted β	Adjusted[a] B [95% CI]	Adjusted[a] β
	South West	158 (15)	0.161 [−0.048, 0.371]	0.065	0.150 [−0.063, 0.362]	0.060	−0.333 [−0.529, −0.138]	−0.144***	−0.264 [−0.466, −0.062]	−0.114*
Education	Secondary education and lower	536 (52)	Reference	Reference	Reference	Reference	Reference	Reference	Reference	Reference
	Higher education and above	489 (48)	−0.044 [−0.153, 0.066]	−0.24	−0.029 [−0.139, 0.081]	−0.016	−0.024 [−0.126, 0.079]	−0.014	−0.038 [−0.142, 0.066]	−0.023
Ethnicity	White	926 (90)	Reference	Reference	Reference	Reference	Reference	Reference	Reference	Reference
	Other	99 (10)	0.015 [−0.171, 0.200]	0.005	0.024 [−0.168, 0.217]	0.008	0.349 [0.177, 0.521]	0.124***	0.284 [0.102, 0.467]	0.101**
Household income	£0 to 25,000	354 (35)	Reference	Reference	Reference	Reference	Reference	Reference	Reference	Reference
	£25,000 to 49,000	313 (30)	−0.094 [−0.229, 0.042]	−0.048	−0.061 [−0.194, 0.073]	−0.031	0.077 [−0.050, 0.204]	0.042	0.090 [−0.037, 0.217]	0.050
	£49,000+	228 (22)	−0.212 [−0.361, −0.064]	−0.099**	−0.155 [−0.305, −0.004]	−0.072*	−0.109 [−0.248, 0.030]	−0.054	−0.101 [−0.244, 0.041]	−0.050
	Rather not say	130 (13)	−0.060 [−0.239, 0.120]	−0.022	−0.089 [−0.265, 0.086]	−0.033	0.016 [−0.152, 0.184]	0.006	0.016 [−0.151, 0.183]	0.006

[a] controlled for sex, age, region, education, ethnicity, and household income

$^{*}p < 0.05$, $^{**}p < 0.01$, $^{***}p < 0.001$

Table 3.14 Flood-related background variables and its associations with factor scores related to NRR security outcomes ($N = 1{,}025$)

Variables	Levels	Number (%)	*Personal impact*				*Community impact*			
			Unadjusted B [95% CI]	Unadjusted β	Adjusted[a] B [95% CI]	Adjusted[a] β	Unadjusted B [95% CI]	Unadjusted β	Adjusted[a] B [95% CI]	Adjusted[a] β
Flood insurance	No	326 (32)	Reference	Reference	Reference	Reference	Reference	Reference	Reference	Reference
	Yes	492 (48)	0.016 [−0.110, 0.141]	0.009	0.121 [−0.005, 0.248]	0.068	−0.127 [−0.243, −0.010]	−0.076*	−0.078 [−0.198, 0.043]	−0.047
	Not sure	207 (20)	0.062 [−0.094, 0.218]	0.028	0.067 [−0.087, 0.221]	0.030	−0.114 [−0.259, 0.032]	−0.055	−0.114 [−0.261, 0.032]	−0.055
Length in community	0 to 4 years	281 (27)	Reference	Reference	Reference	Reference	Reference	Reference	Reference	Reference
	5 to 10 years	228 (22)	−0.032 [−0.189, 0.124]	−0.015	−0.019 [−0.174, 0.134]	−0.009	0.082 [−0.064, 0.228]	0.041	0.095 [−0.052, 0.243]	0.048
	11 to 20 years	262 (26)	−0.029 [−0.180, 0.122]	−0.014	−0.001 [−0.151, 0.149]	−0.001	−0.017 [−0.158, 0.123]	−0.009	−0.011 [−0.154, 0.131]	−0.006
	21 + years	254 (25)	0.008 [−0.144, 0.160]	0.004	0.034 [−0.121, 0.188]	0.016	0.036 [−0.106, 0.178]	0.019	0.040 [−0.106, 0.187]	0.021
Previous flooding	No	913 (89)	Reference	Reference	Reference	Reference	Reference	Reference	Reference	Reference
	Yes	112 (11)	−0.043 [−0.219, 0.132]	−0.015	−0.005 [−0.178, 0.168]	−0.002	0.119 [−0.045, 0.283]	0.044	0.119 [−0.044, 0.283]	0.045
Perceived flood risk	No	730 (71)	Reference	Reference	Reference	Reference	Reference	Reference	Reference	Reference
	Yes	200 (20)	0.005 [−0.135, 0.145]	0.002	0.052 [−0.085, 0.190]	0.023	−0.109 [−0.240, 0.022]	−0.052	−0.072 [−0.203, 0.059]	−0.034

(continued)

Table 3.14 (continued)

Variables	Levels	Number (%)	*Personal impact*				*Community impact*			
			Unadjusted B [95% CI]	Unadjusted β	Adjusted[a] B [95% CI]	Adjusted[a] β	Unadjusted B [95% CI]	Unadjusted β	Adjusted[a] B [95% CI]	Adjusted[a] β
	Don't know	95 (9)	−0.143 [−0.334, 0.049]	−0.046	−0.185 [−0.374, 0.003]	−0.060	0.064 [−0.114, 0.243]	0.022	0.026 [−0.153, 0.206]	0.009
Perceived belonging to community	Not at all or not very strongly	500 (49)	Reference	Reference	Reference	Reference	Reference	Reference	Reference	Reference
	Fairly or very strongly	525 (51)	−0.030 [−0.140, 0.079]	−0.017	−0.019 [−0.127, 0.089]	−0.011	−0.192 [−0.294, −0.090]	−0.115***	−0.191 [−0.292, −0.089]	−0.114***
Knowing location of local disaster support centre	No, don't' know	406 (40)	Reference	Reference	Reference	Reference	Reference	Reference	Reference	Reference
	Yes	159 (16)	−0.147 [−0.311, 0.017]]	−0.059	−0.088 [−0.250, 0.073]	−0.036	−0.224 [−0.376, −0.072]	−0.097**	−0.208 [−0.361, −0.056]	−0.090**
	No, but have guess	460 (45)	−0.056 [−0.175, 0.063]	−0.031	0.003 [−0.116, 0.121]	0.002	−0.228 [−0.338, −0.117]	−0.136***	−0.184 [−0.296, −0.072]]	−0.110***

[a] controlled for sex, age, region, education, ethnicity, and household income
$*p < 0.05$, $**p < 0.01$, $***p < 0.001$

Table 3.15 Results of the linear regression analyses between the respective factor scores of the elements and the NRR security outcomes

Variables	*Personal impact*				*Community impact*			
	Unadjusted B [95% CI]	Unadjusted β	Adjusted[a] B [95% CI]	Adjusted[a] β	Unadjusted B [95% CI]	Unadjusted β	Adjusted[a] B [95% CI]	Adjusted[a] β
Strong fellowship	0.056 [−0.005, 0.116]	0.056	0.064 [0.004, 0.125]	0.065*	−0.205 [−0.260, −0.150]	−0.223***	−0.190 [−0.246, −0.133]	−0.206***
Trust in system	−0.029 [−0.091, 0.034]	−0.028	−0.023 [−0.085, 0.038]	−0.023	−0.141 [−0.199, −0.083]	−0.148***	−0.141 [−0.198, −0.083]	−0.148***
Memories of disasters	−0.075 [−0.137, −0.012]	−0.073*	−0.077 [−0.139, −0.015]	−0.075*	0.126 [0.067, 0.184]	0.131***	0.112 [0.053, 0.171]	0.117***

[a] controlled for sex, age, region, education, ethnicity, and household income
$^{*}p < 0.05$, $^{**}p < 0.01$, $^{***}p < 0.001$

3.4 Discussion

This Chapter explored how the elements of community resilience would interact with the likely outcomes of flooding perceived by people living with high flood risk. Over half of the survey participants viewed their community as having high-quality health services available, enough local knowledge to cope with a flood for a few days, and community members who would help each other, even if they were not affected. Over half of the participants also said they did not know who would be in charge if an emergency occurred in the community. Drawing from the exploratory factor analysis, Strong Fellowship, Trust in the System, and Memories of Disasters were the essence of what participants at a high flood-risk thought of when answering questions related to the elements of community resilience. The NRR security outcomes were perceived as a likely negative personal or community impact caused by a flood. The respective factor scores of the elements of community resilience and the NRR security outcomes were tested and measured for association through quantitative analysis. Exploring these associations further will help the understanding and measuring of community resilience and its application by quantifying the possible effect in practice.

For example, the survey participants who believed they were highly involved in their community felt that a likely flood would lead to a bigger burden on them. This burden may be due to the dual efforts of having to help others in their neighborhood while dealing with the participant's own flooding impacts. Some participants in the focus groups from Chap. 2 expressed concerns about being a good neighbor and trying to cope with the impact of the flood on their own households. Increased burden resulting from helping others and the possible creation of isolation was also seen in the qualitative interviews with disaster response workers by Ulman [45]. A specific study will need to be carried out to understand better the actual effect shared by focus group participants from Chap. 2 and the disaster response workers in the study by Ulman [45].

Based on the total variance explained by the two factors related to the NRR security outcomes, the survey results also highlighted a more significant concern among participants for the likely impact a flood would have on them rather than the community. This concern may suggest that a greater emphasis may be needed on individual resilience within the growing literature, e.g., Bonanno et al. [4], Boscarino and Adams [5], Caffo and Belaise [8], Dollwet and Reichard [16], Polk [40], and Iacoviello and Charney [24]. However, as Norris et al. [36] inferred, community resilience is more than solely adding up the levels of individual resilience in a community, additional research is needed to learn whether there is an interconnectedness between individual resilience and community resilience, or a minimum required threshold for to help reduce NRR security outcomes.

An association found in this survey was that participants with more memories of disasters were less likely to perceive that a flood would affect them personally but more likely to think a flood would harm the community. Learning and applying lessons from past disasters is an established practice to reduce knowledge gaps within communities and enhance preparedness at an individual and community level [12,

46]. The difference between personal and community effects could be due to the participants having greater confidence in their ability to take more responsibility for preparing against a flood more efficiently in their household than in their community. However, participants with a strong fellowship in the community or greater trust in the system in place were less likely to perceive that a flood would have a community-wide harmful impact. This perception was the same finding for the focus group participants who had higher trust in the system and those in charge. This finding also supports previous studies and reports that found trust to be critical in building community resilience against disasters [1, 2, 14, 34, 44]. Participants who felt that they belonged to their community or had a guess or knew the location of the local disaster support center in their community were less likely to think that a flood would impact the community. This finding suggests that the elements of community networks and relationships, governance and leadership, and local knowledge play an essential role in the participant's perception of a likely flood.

The survey participants understood the elements of community resilience as overlapping concepts that fit into three factors, which differed from the focus group participants. This survey also identified how flood-risk participants understood NRR security outcomes in a fictional flooding scenario. Although most survey participants had little or no prior flooding experience, they understood NRR security outcomes as two impacts: adverse effects at a personal level and a community level. For practitioners, this understanding can better inform how to apply the elements of community resilience in high flood-risk areas. A practitioner should make sure to focus on two levels, especially if many of the people in the community have no direct flooding experience: (1) creating more memories of disasters to reduce the likely personal impact caused by floods, and (2) improving the community's fellowship, trust in the system, and memories of disasters.

It is crucial to consider the differences in how the flood-affected participants in Chap. 2 and the high flood-risk respondents in this chapter made sense of this unfamiliar information about community resilience against flooding. Having experienced flooding, the focus group participants were well-informed and thought deeply about how to better their community against flooding. Despite being a cross-sectional survey and relying on an exploratory factor analysis, the participants with high flood risk had less crystallized thoughts on the elements of community resilience based on the factors found. They even saw how these factors were associated with flood impacts on two levels: personal and community. Additionally, the differences identified between the participants in Chap. 2 and in this Chapter (Chap. 3) could be similarly seen in the people who had prior experience with earthquakes in the study by Gruev-Vintila and Rouquette [22], where the differences between those who had or had not experienced an earthquake contributed to the different levels of recognition and practical application of risk-related information. The findings of both chapters did show that the flood-affected and flood-risk participants understood community resilience differently, and practitioners may want to follow the suggestion by Gruev-Vintila and Rouquette [22] in increasing the efficiency and effectiveness of a prevention campaign between populations, such as using a collective risk-mitigation approach in the communities.

3.4.1 *Meaning of 'Community'*

Everyone has a different understanding of the community. It is not surprising that participants most commonly endorsed immediate neighbors, local friends, wider neighbors, and immediate family as comprising part of their community. Locality as part of the definition of community has been previously discussed and addressed in the literature and by governments [6, 25, 27, 33, 35]. Generally, the terms locality and community have been used interchangeably and assumed to be synonyms throughout the survey participants' responses, as the majority agreed to the vast choices given in the question about the meaning of community. Although a minor agreement of the survey participants' understanding of community included online contacts, people in the same country, and religious or faith-based groups, these choices, which more than 25% of participants agreed with, represent the notion that a community does not necessarily have to be physically located near a person. However, many of the survey participants agreed that people located near them aligned better with their meaning of community than their online contacts.

The findings on the meaning of community are interesting as it may partly explain the various ambiguous understandings of community resilience. If the meaning of 'community' varies in its understanding, then how can community resilience have a unifying one? The assumptions made in interpreting 'community' may be one of the barriers to reaching a consensus on community resilience. For example, ten different choices were given to the participants when asked about their definition of community. At least 25% of the survey participants agreed or strongly agreed with each option, showing a varied interpretation of community. Due to this diverse understanding of community, perhaps the approach of a catch-all terminology may be more appropriate for the concept of community resilience if a formal definition is needed.

3.4.2 *Convergence of Focus Group Discussions and Survey Results*

A benefit of Chap. 2 was that participants added specific details to each element of community resilience. These details were also found in other studies. The focus group participants explained their pre- and post-flood perceptions of risk and preparedness. Some focus group participants discussed the importance of a feeling of belonging to the community. Among older adults affected by the Great East Japan Earthquake, Kiyota et al. [28] found their participant's sense of belonging and trust was essential to the community's bonding and levels of social network and participation. Both, a sense of belonging and trust, would contribute to enhancing the element of community networks and relationships.

The focus group participants also discussed their issues with flood insurance and their lack of confidence in insurance companies after their poor experience with them.

Carroll et al. [9] found participants sharing similar disputes about flood insurance. In their qualitative study focusing on flooding in northwest England, participants stated that they had higher levels of psychological stress because of the flood insurance companies. This finding by Carroll et al. [9] echoed the conclusions of the focus group discussions in Chap. 2. Regarding a lack of confidence in flood insurance, publications by Lamond and Proverbs [30], Treby et al. [43], and Harries [23] provided even greater detail on the matter with policy implications on the insurance risk transfer and how people's poor experience with flood insurance damaged their confidence in having insurance in the UK. Lastly, the focus group participants from Chap. 2 shared their feelings of heightened engagement with their own flood preparedness. Keeping updated emergency plans and restocking household emergency supplies, some groups said these were even done at a community level as well as the personal level. Based on a longitudinal study consisting of people living in London, UK, following the London bombings of July 7th, 2005, Page et al. [37] found that individual preparedness and specific emergency plans increased seven to eight months after the bombings. The details found from the focus groups aligned with prior studies. These details in the elements of community resilience could help enhance communities against flooding at a practical level. However, would the elements enhance communities against flooding?

From this chapter's survey results, the high flood-risk participants understood the elements of community resilience as three distinct positive notions: strong fellowship in the community, trust in the system, and more memories of disasters. The details found in the focus groups appear to diverge from the survey findings over the distinction between elements. The flood-risk participants saw the elements of community resilience as not distinct elements but a blending of several elements that encompasses positive ideas that will help against the likely impacts caused by flooding. However, these ideas of strong fellowship, trust, and memories of disasters were too divergent from the focus group findings in Chap. 2. Many focus groups confirmed similar beliefs about the importance of those three ideas as positive actions that can help them reduce their flood impacts. For example, the strong fellowship was seen in the groups when discussing the connectedness and cohesion of the community after the flood. Trust was mentioned in various elements in the focus groups, such as governance and leadership, and communication. The discussions about local knowledge, preparedness, and mental outlook align with this chapter's findings on memories of disasters. Even though the focus group participants provided specific details while the flood-risk participants gave broad attributes, both results confirm the hypothesis that applying elements of community resilience can reduce the flood impact in communities.

3.4.3 Limitations

There are several limitations to consider with the survey presented in this chapter. First, access was only given to a single public database. Still, it included a descriptive summary of the methodology for the flood estimation, and the database was developed and supported by the EA.

Second, the participants in the survey may not live in homes that could even be flooded (e.g., flats in a multi-story building), despite being in a postcode determined by the EA to be at high risk for flooding. The public database did not mention whether vertical flood risk or a property-level assessment were accounted for. The flood risk was assumed to be equal for all homes within each postcode identified from the database. According to the methods of determining the flood risk for the database, the authors accounted for the depth of floodwater and whether floodwater may cause damage or disruption to homes, roads, or other infrastructure in the given postcode [19]. Not having additional detailed information on vertical flood risk and specific property level assessment for each participant could mean that the actual flood risk of the survey participants may be lower than the designation of high risk found in the public database. The real flooding risk for the participants may be even closer to the participant's perceived flood risk, which a majority said was none.

Third, questions designed to represent each element of community resilience and NRR security outcome may not be reliable to measure each element or outcome accurately. Each question was based on the scientific literature or results from the focus group discussions. The survey was tested through piloting with people who were recently affected by flooding before being used in the larger population. The feedback from the pilot gave insight on questions that were not understood well and needed to be re-worded or deleted.

Fourth, using a fictional flooding situation may not help to imagine what the actual findings of the survey would be if it were carried out before and after a disaster. Using a pre-test and post-test design would provide a clearer picture of the actual associations seen between the elements of community resilience and the NRR security outcomes. Still, the approach and findings of the survey have left a solid foundation to build future work upon, especially if validated scales are developed for each element and outcome as it relates to flooding.

Fifth, the factor analysis performed was exploratory, and the statements did not load uniquely to one factor. The factors were presented by assigning statements by their largest loading value to a specific factor. This binary approach of all or nothing was not used in the statistics and calculations, which accounted for each statement not loading uniquely. These calculations could mean that the factors found in the survey could be broader in scope than only including strong fellowship, trust, and memories of disasters, if accounting for the influence of the overlapped statements that were assumed to be excluded. Nevertheless, the associations were accurate due to not taking a binary assumption as used to label the theme of each factor.

Sixth, the associations found can hold true either way, and the elements of community resilience could be an outcome due to a flood. The assumption made in this survey was that each element of community resilience would act as a predictor of a possible NRR security outcome caused by a flood. This assumption may not be accurate. Specifically, a choice was needed to be made between whether community resilience is best understood as a static idea or a process of change. If the former, then developing elements into predictors would be rooted better to test the association with NRR security outcomes (the approach used in this survey). However, if community resilience was chosen as the latter idea, then further development of these elements as to how to measure them in the cycle of a disaster, as not all elements may be present before a flood. Some elements may arise due to the flood. Thus, measuring the process of change of the elements in a non-flooded population would be complicated and may have a greater likelihood of flaws in the survey methodology. Despite these choices, this survey illuminated the additional complexities within community resilience as it applies to disasters and in its constituent elements as it applies to flooding. Both need further evidence to know which approach is closer to how the concept and elements are embodied in communities during a real disaster.

3.4.4 Implications

Several implications were found and should be considered from the survey findings. First, Strong Fellowship was understood as the participant's belief that 'my community can handle flooding' and a strong sense of mutual cooperation and fellowship in the community. There was a discussion about this symbiotic and positive feeling in the focus groups. The feeling of strong fellowship was also seen as something that did not last long once the actual floods occurred. However, could there be a way for practitioners to bring congruence between this belief and apply it to an actual flood? Based on the review given by Patel et al. [38], the feeling of strong fellowship was mentioned as part of the definition of resilience. For example, when authors discussed learning lessons from previous disasters, resilience was mentioned alongside improving the community's sense of fellowship [2, 15, 41], and even established within proposed frameworks of community resilience [10, 36]. Most of the applications of this intertwining of resilience and strong fellowship included empowerment techniques in the community and building a community's social capital to better protect the community against a disaster.

Second, Trust in the System encapsulated the idea that community leaders, emergency services, and other officials would aid the participant's community if a disaster occurred. A large proportion of the high flood-risk participants strongly agreed or agreed with most statements in this factor and believed that necessary aid would actualize if a real flood happened to them. As seen from the focus group discussions, many flood-affected participants held the same belief until their trust was broken when no help or assistance arrived after the flood. The lack of support and aid led many focus group participants to feel let down and isolated from the system meant to help them. With respect to the literature related to community resilience, the concept of trust was found in the definitions by Coles and Buckle [11] and Ganor and Ben-Lavy [21]. Both publications emphasized the importance of establishing and growing credibility between the leadership and community members and acting on mutual support against a disaster.

Third, Memories of Disasters included both sharing lessons to cope with flooding and assuming that the community would still be traumatized by previous disasters. This finding adds strength to the ongoing call among researchers and practitioners for sharing and maintaining knowledge exchange and transfer of lessons identified and outcomes for action within communities. In the emergency management literature, this knowledge exchange and transfer were described by Comfort and Okada [12] as "knowledge commons," and the authors stressed the importance of developing such function before a disaster so leaders making the decisions during a disaster can easily incorporate the knowledge of the past. During disasters, making the same mistakes could be avoided if knowledge sharing is a common practice within communities. Additionally, this factor suggests that the value of the element of mental outlook could have great practical importance moving forward. Learning from previous lessons establishes a growth mindset among the community, which aligns with the element of mental outlook.

Further research will need to be conducted to confirm the high flood-risk participants' perceived impacts based on the NRR security outcomes. Still, if consistent, then the factors found for NRR security outcomes have important implications, especially in communicating flood risks to the general public living in high-risk areas. For example, it may not be enough to inform the public about the risk numbers or even provide flood risk maps during the discussion. The survey participants were mainly concerned about two harmful impacts of likely flooding. The survey participants were concerned about the damage on personal and community levels. Future emphasis on flood preparation should complement and explain the details for the high-risk areas at these levels.

Exploring the interworking of elements of community resilience, especially the interactions and overlaps between them, will need further research. For example, the statements loaded onto Strong Fellowship included statements intended to measure five elements of community resilience: local knowledge, communication, community networks and relationships, mental outlook, and health. This is like the other factors found among the elements of community resilience. Could this mean that elements of community resilience were too broadly construed and ambiguous like the definitions of community resilience? This is not necessarily so; there could be more at play when observing and identifying the elements. For example, the amount of flood experience differed between the participants of the focus groups and the survey. A key distinction of attaining first-hand knowledge among the participants may be why the survey participants did not distinctly see all the elements of community resilience as separate items.

3.5 Conclusion

In this chapter, a cross-sectional online survey was carried out among participants who live in high-risk areas for flooding in the UK ($N = 1{,}025$). This survey was split into four sections: (a) participants' understanding of community, risk of flooding, and household emergency preparedness kits; (b) participants' agreement with statements related to elements of community resilience; (c) participants' agreement with statements related to likely psychological impact, economic harm, and social disruption in the event of a flood; and (d) participants' demographic characteristics. According to an exploratory factor analysis, participants interpreted elements of community resilience as three attributes of a community (Strong Fellowship, Trust in the System, and Memories of Disasters) and harmful security and health outcomes (NRR security outcomes) as two flood impact levels (personal and community). Participants with a strong fellowship in the community or fewer memories of disasters perceived that a potential flood would have a personal impact ($p < 0.05$). Participants with a weak fellowship in the community, more memories of disasters, or a lack of trust in the system and leadership thought a likely flood would have a community impact ($p < 0.001$). Participants who felt that they belonged to their community or had a guess or knew the local disaster support center's location in their community thought a flood would have no community impact ($p < 0.001$). The findings did not suggest the existence of a hierarchy of priorities among elements of community resilience about flooding due to the elements overlapping with each other in the survey results.

References

1. Abramson DM, Grattan LM, Mayer B, Colten CE, Arosemena FA, Bedimo-Rung A et al (2015) The resilience activation framework: a conceptual model of how access to social resources promotes adaptation and rapid recovery in post-disaster settings. J Behav Health Serv Res 42(1):42–57. https://doi.org/10.1007/s11414-014-9410-2
2. Aldrich DP, Meyer MA (2015) Social capital and community resilience. Am Behav Sci 59(2):254–269. https://doi.org/10.1177/0002764214550299
3. BBC (2016) Storms Desmond and Eva flooded 16,000 homes in England. http://www.bbc.co.uk/news/uk-35235502
4. Bonanno GA, Galea S, Bucciarelli A, Vlahov D (2006) Psychological resilience after disaster: New York City in the aftermath of the September 11th terrorist attack. Psychol Sci 17(3):181–186. https://doi.org/10.1111/j.1467-9280.2006.01682.x
5. Boscarino JA, Adams RE (2009) PTSD onset and course following the World Trade Center disaster: findings and implications for future research. Soc Psychiatry Psychiatr Epidemiol 44(10):887–898. https://doi.org/10.1007/s00127-009-0011-y
6. Cabinet Office (2014) Community development framework. Civil Service Reform. https://www.gov.uk/guidance/community-development-framework
7. Cabinet Office (2015) National risk register for civil emergencies-2015 edition. London, UK. https://www.gov.uk/government/uploads/system/uploads/attachment_data/file/419549/20150331_2015-NRR-WA_Final.pdf

8. Caffo E, Belaise C (2003) Psychological aspects of traumatic injury in children and adolescents. Child Adolesc Psychiatr Clin N Am 12(3):493–535. https://doi.org/10.1016/S1056-4993(03)00004-X
9. Carroll B, Balogh R, Morbey H, Araoz G (2010) Health and social impacts of a flood disaster: responding to needs and implications for practice. Disasters 34(4):1045–1063. https://doi.org/10.1111/j.1467-7717.2010.01182.x
10. Chandra A, Acosta J, Stern S, Uscher-Pines L, Williams M, Yeung D et al (2011) *Building community resilience to disasters: A way forward to enhance national health security (technical report)*. Arlington, VA: RAND Corporation Retrieved from https://www.rand.org/content/dam/rand/pubs/technical_reports/2011/RAND_TR915.pdf
11. Coles E, Buckle P (2004) Developing community resilience as a foundation for effective disaster recovery. Aust J Emergency Manag 19(4), 6. Retrieved from https://search.informit.com.au/documentSummary;dn=375435145094637;res=ielhss
12. Comfort LK, Okada A (2013) Emergent leadership in extreme events: a knowledge commons for sustainable communities. Int Rev Public Administr 18(1):61–77. https://doi.org/10.1080/12294659.2013.10805240
13. Coulston JE, Deeny P (2010) Prior exposure to major flooding increases individual preparedness in high-risk populations. Prehosp Disaster Med 25(04):289–295. https://doi.org/10.1017/S1049023X00008219
14. Cox RS, Hamlen M (2015) Community disaster resilience and the rural resilience index. Am Behav Sci 59(2):220–237. https://doi.org/10.1177/0002764214550297
15. Dawes SS, Cresswell AM, Cahan BB (2004) Learning from crisis: lessons in human and information infrastructure from the world trade center response. Soc Sci Comput Rev 22(1):52–66. https://doi.org/10.1177/0894439303259887
16. Dollwet M, Reichard R (2013) Assessing cross-cultural skills: validation of a new measure of cross-cultural psychological capital. Int J Human Resource Manag (ahead-of-print), 1–28
17. Environment Agency (2015a) Risk of flooding from rivers and sea: flood map. http://watermaps.environment-agency.gov.uk/wiyby/wiyby.aspx?topic=floodmap&scale=7&ep=map&layerGroups=default&lang=_e&y=181500&x=531500#x=531502&y=181583&scale=6
18. Environment Agency (2015b) RoFRS Postcode England v201504. http://www.geostore.com/environment-agency/WebStore?xml=environment-agency/xml/ogcDataDownload.xml
19. Environment Agency (2015c) Risk of flooding from rivers and sea - postcodes in areas at risk: product description. The National Archives, contains public sector information licensed under the Open Government Licence v3.0 Retrieved from http://www.geostore.com/environment-agency/WebStore?xml=environment-agency/xml/ogcDataDownload.xml
20. Fabrigar LR, Wegener DT, MacCallum RC, Strahan EJ (1999) Evaluating the use of exploratory factor analysis in psychological research. Psychol Methods 4(3):272–299. https://doi.org/10.1037/1082-989X.4.3.272
21. Ganor M, Ben-Lavy Y (2003) Community resilience: lessons derived from Gilo under fire. J Jewish Communal Serv 79(2/3):105–108. Retrieved from http://www.bjpa.org/Publications/details.cfm?PublicationID=1223
22. Gruev-Vintila A, Rouquette ML (2007) Social thinking about collective risk: how do risk-related practice and personal involvement impact its social representations? J Risk Res 10(4):555–581. https://doi.org/10.1080/13669870701338064
23. Harries T (2012) The anticipated emotional consequences of adaptive behaviour—impacts on the take-up of household flood-protection measures. Environ Plan A 44(3):649–668. https://doi.org/10.1068/a43612
24. Iacoviello BM, Charney DS (2014) Psychosocial facets of resilience: implications for preventing posttrauma psychopathology, treating trauma survivors, and enhancing community resilience. Europ J Psychotraumatology 5. https://doi.org/10.3402/ejpt.v5.23970
25. Johnson D, Headey B, Jensen B (2005) Communities, social capital and public policy: literature review. (26). Canberra: Commonwealth of Australia Retrieved from https://www.dss.gov.au/sites/default/files/documents/05_2012/prp_no_26.pdf

26. Kaniasty K (2012) Predicting social psychological well-being following trauma: the role of postdisaster social support. Psychol Trauma Theory Res Pract Policy 4(1):22–33. https://doi.org/10.1037/a0021412
27. Kearns A, Parkinson M (2001) The significance of neighbourhood. Urban Studies 38(12):2103–2110. https://doi.org/10.1080/00420980120087063
28. Kiyota E, Tanaka Y, Arnold M, Aldrich DP (2015) Elders leading the way to resilience. Washington, DC
29. Kroenke K, Spitzer RL, Williams JB, Löwe B (2009) An ultra-brief screening scale for anxiety and depression: the PHQ–4. Psychosomatics 50(6):613–621
30. Lamond JE, Proverbs DG, Hammond FN (2009) Accessibility of flood risk insurance in the UK: confusion, competition and complacency. J Risk Res 12(6):825–841. https://doi.org/10.1080/13669870902768614
31. Lochner K, Kawachi I, Kennedy BP (1999) Social capital: a guide to its measurement. Health Place 5(4):259–270
32. Löwe B, Wahl I, Rose M, Spitzer C, Glaesmer H, Wingenfeld K et al (2010) A 4-item measure of depression and anxiety: validation and standardization of the Patient Health Questionnaire-4 (PHQ-4) in the general population. J Affect Disord 122(1):86–95. https://doi.org/10.1016/j.jad.2009.06.019
33. MacQueen KM, McLellan E, Metzger DS, Kegeles S, Strauss RP, Scotti R et al (2001) What Is community? An evidence-based definition for participatory public health. Am J Public Health 91(12):1929–1938. https://doi.org/10.2105/AJPH.91.12.1929
34. Mayunga JS (2007) Understanding and applying the concept of community disaster resilience: a capital-based approach. Summer Acad Soc Vulnerability Resilience Build 1:16
35. Monk S, Clark A, Tang CPY (2011) Mixed communities literature review. Edinburgh, Scottish Government Retrieved from http://www.cchpr.landecon.cam.ac.uk/Projects/Start-Year/2010/Mixed-Communities-Review/Report-to-Scottish-Government-Communities-Analytical-Services/Mixed-Communities-Literature-Review
36. Norris FH, Stevens SP, Pfefferbaum B, Wyche KF, Pfefferbaum RL (2008) Community resilience as a metaphor, theory, set of capacities, and strategy for disaster readiness. Am J Community Psychol 41(1–2):127–150. https://doi.org/10.1007/s10464-007-9156-6
37. Page L, Rubin J, Amlôt R, Simpson J, Wessely S (2008) Are londoners prepared for an emergency? A longitudinal study following the London bombings. Biosecur Bioterror 6(4):309–319. https://doi.org/10.1089/bsp.2008.0043
38. Patel SS, Rogers MB, Amlôt R, Rubin GJ (2017) What do we mean by 'community resilience'? A systematic literature review of how it is defined in the literature. PLoS currents 9
39. Pfefferbaum RL, Pfefferbaum B, Van Horn RL (2011) Communities advancing resilience toolkit (CART): the CART integrated system. Oklahoma City, OK. Retrieved from https://www.oumedicine.com/docs/ad-psychiatry-workfiles/cart_online-final_042012.pdf?sfvrsn=2
40. Polk LV (1997) Toward a middle-range theory of resilience. Adv Nurs Sci 19(3):1–13. Retrieved from http://www.ncbi.nlm.nih.gov/pubmed/9055026
41. Shirani F (2011) The centrality of social capital: concepts of community in the era of the big society [working paper]. School of Social Sciences Working Papers Series, vol 146. Cardiff, UK: Cardiff University Retrieved from http://orca.cf.ac.uk/78205/1/wp146.pdf.
42. Statistical and Technical Team (2001) A practical guide to sampling. UK National Audit Office Retrieved from https://www.nao.org.uk/report/sampling-guide/
43. Treby EJ, Clark MJ, Priest SJ (2006) Confronting flood risk: implications for insurance and risk transfer. J Environ Manage 81(4):351–359. https://doi.org/10.1016/j.jenvman.2005.11.010
44. Twigg J (2009) Characteristics of a disaster-resilient community: a guidance note (version 2). Teddington, UK: DFID Disaster Risk Reduction NGO Interagency Group Retrieved from http://discovery.ucl.ac.uk/1346086/.
45. Ulman KH (2008) Helping the helpers: groups as an antidote to the isolation of mental health disaster response workers. Group 32(3):209–221. Retrieved from https://www.jstor.org/stable/41719201
46. World Vision, & Rumsey R (2014) Disaster risk reduction and community resilience: key concepts and practices. World Vision Retrieved from https://www.yumpu.com/s/ONalu7Z871dYouUC

Chapter 4
Revisiting How Community Resilience Relates to the Security and Health Harm Caused by Severe Flooding: An Analytical Literature Review

Abstract This Chapter reviews the scientific literature to examine previously established associations of elements common to community resilience and the security and health harm caused by severe flooding. Using an analytical narrative review approach, 16 studies tested the association between the elements of community resilience and a community's psychological, social, and economic impacts caused by severe flooding. Most included publications (84%) assessed the association of community resilience with psychological impact, while fewer assessed social disruption (31%) or economic harm (21%). Past studies' gaps were critically reviewed to understand better how community resilience connects to positive and negative outcomes seen after severe flooding. Future studies can clarify whether the community resilience elements are protective factors against harmful outcomes and when best to apply the concept in the disaster cycle.

Keywords Community resilience · Associations · Flooding · Public health · Literature review · Psychological impact · Economic harm · Social disruption · Disasters · Disaster risk reduction · Disaster preparedness

4.1 Background

As mentioned in Chap. 1, current definitions of community resilience vary by interpretation, such as being described as a process of change, a capacity-building ability, a trajectory of functioning, or even an adaptation to life after a disaster. The systematic review of the definitions of community resilience presented by Patel et al. [57] failed to find an existing, common understanding of the concept across many publications and reports. Despite this ambiguity, Patel et al. [57] did identify nine elements that were common to the existing definitions. These nine elements were broadly categorized as: local knowledge, community networks and relationships, communication, health, governance and leadership, resources, economic investment, preparedness, and mental outlook.

In previous chapters, these elements were investigated further about their usefulness in practice. Both qualitative and quantitative evidence were gathered from the

S. S. Patel, *Community Resilience When Disaster Strikes*, Advanced Sciences and Technologies for Security Applications,
https://doi.org/10.1007/978-3-031-07992-4_4

lived experiences of those affected by flooding and those living in high-risk flood areas. Several participants discussed the positive and negative outcomes associated with their community resilience as it applied to a severe flood. The implications of these findings are important as they can inspire evidence-based policy and are also helpful for academic professionals and community practitioners. Turning these findings into lessons learned makes for a more palatable use of the literature, which can inspire future directions, as will be done in the succeeding chapter. This Chapter revisits the past studies on severe flooding and discovers whether NRR security outcomes (psychological impact, economic harm, and social disruption) [11] were measured and connected to any of the nine community resilience elements.

4.2 Methods

A literature review was completed to examine the relationship between the elements of community resilience and the NRR security outcomes discussed in Chap. 1. This review only included peer-reviewed scientific studies that included people who had experienced flooding, like participants from Chap. 2. Papers must have reported at least one statistical association between study variables that closely matched the descriptions of an element of community resilience and an NRR security outcome to be part of this review.

4.2.1 Search Strategy

Papers were found through keyword searches using OvidSP on MEDLINE and PsycInfo databases with a cut-off publication date of October 11, 2015. The search strategy included terms relating to flooding in conjunction with words pertaining to psychological impact, economic harm, and social disruption (see Table 4.1). Because 'impact,' 'harm,' and 'social' are common terms, they were not included in the search, but other keywords for each NRR security outcome were included. Database filters in OvidSP were used to exclude non-human and review studies for this search.

Table 4.1 Keywords used for database searches

Keywords for MEDLINE searches
Flood* AND [**Psych***, anxiety, depression, PTSD, trauma, mental, stress, well-being, wellbeing, OR **Econo***, mone*, OR **Disrupt***, losses]

4.2.2 Inclusion Criteria

Papers were included if they were: (1) written in English; (2) published in a peer-reviewed journal; (3) included original data collected from people who were flooded; (4) quantitatively measured at least one NRR security outcome such as psychological impact, economic harm or social disruption caused by a flood; (5) quantitatively measured a variable or a factor which was broadly consistent with an element of community resilience; and (6) reported at least one association between an element of community resilience and a NRR security outcome.

Studies with original data were defined as studies that included data directly collected from participants affected by flooding. Papers were accepted if the publication discussed climate change or other types of disasters, as long as quantitative data on flooding and the consequences of flooding on affected people were included as part of the study. For example, water from hurricanes, storms, and tsunamis can lead to flooding after its initial onset; thus, papers related to such disasters were included only if flood-related data were clearly labeled. For example, Davis et al. [16], Norris et al. [54], and Weems et al. [81] included participants who were affected by Hurricane Katrina. Still, only the former publication was included in this review because the authors reported associations based on participants who were flooded by the hurricane. Papers that broadly construed flooding as part of climate change were excluded, for example, DeBono et al. [17] Hartz et al. [27], Kjellstrom and McMichael [32], and Warren et al. [79]. Any studies solely reporting qualitative data were excluded as well.

4.2.3 Selection Process

All citations found by the search were imported to EndNote (Version 7.7.1). Publications were initially screened by reading the title and abstract to remove any duplicates, non-English papers, or papers that did not discuss flooding or report an association between a NRR security outcome and an element of community resilience. Full texts of the remaining articles were downloaded and read. A request was made via interlibrary loan for those not found online for download.

4.2.4 Identifying and Labeling the Elements of Community Resilience and the NRR Security Outcomes

Systematically and consistently labeling elements of community resilience and NRR security outcomes could have been challenging because the original descriptions were broad, but there was some overlap between definitions. A more specific range of descriptions was created for each element of community resilience and NRR security outcome so that the studies' variables could be consistently marked and categorized. To be included as an NRR security outcome, a variable had to be measured and attributed to flooding. The outcomes of **social disruption** included any consequences that negatively interrupted daily life, from loss of access to social aid to perceived isolation. **Psychological impact** had been marked with any outcomes that measured the mental health effects caused by the flooding. Some studies, these include the results from the 12-item short-form health survey [77]. An overlap could have occurred between psychological impact and the elements of community resilience of health and mental outlook. The distinguishing quality in marking was that if a measure assessed post-flooding status, it was counted as an outcome of psychological impact. **Economic harm** included variables that measured the flood impact on the community's economy and the income of the individual households within the affected community. For example, studies that looked at the cost of repairs and personal losses at an individual level were accepted. Similar to the marking of psychological impact, the distinguishing quality between economic harm and economic investment was if a measure described a post-flood financial harm caused by flooding, then it was the former.

Variables matching the elements of community resilience could be measured in pre- or post-flooding timeframe, but all variables marked were treated as factors that could influence a NRR security outcome. **Local knowledge** included any variables measuring information and education in relation to flooding. This could range from the number of people who participated in flood safety courses to individual flood awareness scores. Measures and questions related to **community networks and relationships** included any information on the links within and between communities and their members. This could range from the level of social capital in a community to the number of people helping their neighbors.

Communication included any measure related to risk warnings, crisis requests, and messages for action. The messages could happen before, during, and after a flood, such as a request for evacuation. **Health** included health status, physical and mental health indicators, and assessments of quality and delivery of health services.

Governance and leadership included variables that assessed the decisions and actions provided by the community leadership and the quality and delivery of essential (non-health-related) services. This also included the processes and length of the evacuation, relocation, and the overall sense of getting back to normal. **Resources** were marked as essential supplies (except actual money) needed for preparedness and response. This could be the number of households in the community with a specific resource or at least access to such resource regardless of whether it was used during the flood. **Preparedness** included actions taken before a flood designed to mitigate the effects of a flood. To distinguish between and reduce the overlap between resources and preparedness, variables were seen as preparedness if actions occurred before a flood. In contrast, resources were marked if access to or presence of a supply designed to mitigate the effects of a flood. For example, if a variable measured the stockpile of sandbags, then it would be a resource. If a variable measured the number of people using sandbags, it would be marked as preparedness.

Economic investment included anything related to the economy of the community and its individual households, such as income level and costs required for flood protection, including insurance and even investment made before the flood, i.e., house value. **Mental outlook** included any measures that captured the participant's attitude and coping skills in relation to a flood. An assumption was made that this should consist of variables measuring previous anxiety about flooding and prior flood experience. It may relate more closely to long-lasting fear than a specific physical and mental health indicator that was marked as health.

4.2.5 Analysis

From each publication included in the review, the following information was extracted: study design, period of data collection, flood occurrence, age and gender of a sample, the sample size used in the analysis, measures identified as NRR security outcomes (psychological impact, economic harm, social disruption), and measures identified as variables related to the elements of community resilience. Narrative summaries were created to critically appraise each study and each study's methodology, data collection, analysis, and interpretation of results. Based on these summaries, evidence tables were created for each NRR security outcome and presented in the results section.

4.3 Results

4.3.1 Overall Results

Of 667 publications found in the search, 512 were removed during the screening process. Nineteen publications were included in this review (Fig. 4.1). Among these publications, seven used participants from the United States, six from the United Kingdom, two from Poland, two from Australia, and one each from China and Mexico. The overall details on methodology, data collection, and analysis for each study ranged from very little to enough to determine the quality of the study. Table 4.2 provides methodological details for each study. In this table, measures that were seen as elements of community resilience are in bold if a significant association was found with a NRR security outcome. The following sections provide a detailed narrative review of the specifics of each study, organized by NRR security outcome. This is to identify the reported associations between the elements of community resilience and NRR security outcomes, marked in brackets and italics when appropriate.

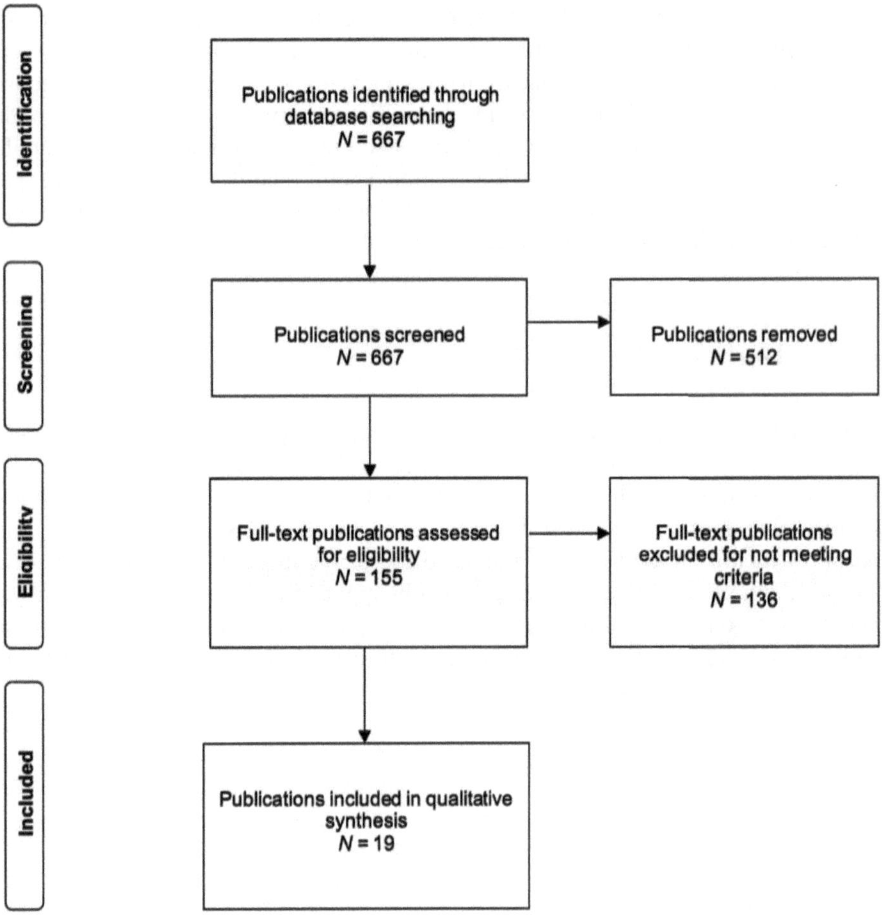

Fig. 4.1 A flow diagram of the selection process

4.3.2 *Outcome of Psychological Impact*

Findings on the associations between psychological impact and an element of community resilience were found in 16 publications. Most papers were cross-sectional studies (85%) and carried out in the UK or USA (69%). Overall, the previous studies show that the participants with existing health problems or concerns, who evacuated, or who reported anxiety from a prior flood experience tend to experience worse psychological impact from a flood. Taking steps to improve preparedness, such as those based on a flood warning, tends to reduce the psychological impact, but not many studies confirm this improvement. Coping strategies, such as emotional coping and having a good physical health status, seem to help reduce the likely impact at an individual level. Table 4.3 captures the evidence of the associations found in each study. These studies are also described in detail in the following sections.

Table 4.2 Overview of the studies included in this review

Reference(s)	Design, country, and period of data collection	Disaster	Age (% or mean ± SD)/gender (%)	Sample size of analysis	Measures of NRR outcomes	Measures of elements Bold if significant association with NRR outcome	Limitations
Alderman et al. [4]	Cross-sectional survey, Australia, 2011	Summer Floods of 2011	18–24: 7% 25–44: 28% 45–64: 43.6% ≥65: 21.3% Male: 43.6% Female: 54.1%	960	Psychological distress—short version Kessler 10 & 6	**Existing health problems** Owning their property	• Selection bias • Different numbers in tables that do not add up to the sample size • Under-represented flooding sample
					PTSD-civilian checklist	**Existing health problems** **Owning their property**	
					Groninger Sleep Quality Scale	**Existing health problems** **Owning their property**	
Bei et al. [5]	Longitudinal prospective study, Australia, 2010–2011	Floods in early 2011	≤70: 51% >70: 49% Male: 28% Female: 72%	53	PTSD (IES-R) and subscales of Intrusion, avoidance, and hyperarousal	**Evacuation**	• Small and differing sample size • Unknown age and gender percentages for sub-sample • Participants surveyed at different times after the flood (exact timing not given)
				45	PTSD (IES-R)	**Social support received**	
					Depression (CES-D)	Social support received Evacuation	
					Anxiety (GAI)	Social support received	
					Short form -12 health survey, physical health component	**Social support received**	
					Short form-12 health survey, mental health component	Social support received	
					Satisfaction with life scale	**Social support received**	

(continued)

Table 4.2 (continued)

Reference(s)	Design, country, and period of data collection	Disaster	Age (% or mean ± SD)/gender (%)	Sample size of analysis	Measures of NRR outcomes	Measures of elements Bold if significant association with NRR outcome	Limitations
Bernatchez et al. [7], Bokszczanin [8]	Cross-sectional study, Poland, 1999	Summer flood of 1997	$n = 532$ 11–15: 45% 16–21: 55% $n = 533$ Male: 40% Female: 60%	503	Sum of symptoms for Revised Civilian Mississippi PTSD Scale	**Availability of social support from parents**	• Unclear methodological design • Potential recall bias and information bias • Sample size of analysis lack descriptive information (i.e., age, gender, etc.)
Chang [12]	Cross-sectional survey, United Kingdom, not specify	2005 winter floods	39.52 ± 15.76 Male: 38.54% Female: 61.46%	96	Severity of disruptions caused by flood	Community affect (based on Sense of Community Index) **Community cognition (based on Social Life Feeling Scale)** Community behavior (based on Community Participation Scale) **Community cohesion (based on self-perceived changes pre- and post-floods)**	• Unknown collection period • Unknown degree of direct or indirect flood impact felt by participants • Potential recall bias and information bias • No baseline (pre-flood) measures were taken, so change due to the flood may not be accurate • Small sample size and missing input from community members who permanently moved out
Chen et al. [13]	Cross-sectional study, United States, 2006–2007	Floods caused by 2005 Hurricane Katrina	"About 40 years age" (p.327) Male: 54% Female: 46%	113	IES-R PTSD symptoms	Previous traumatic experiences	• Possible language bias • Possible recall bias
					SF36—mental health subscale	Previous traumatic experiences	
					SF36—physical health subscale	**Previous traumatic experiences**	
Collins et al. [14]	Cross-sectional survey, United States, 2010	Floods of 2006	Mean age: 35.675 Male: 39% Female: 61%	475	Mental health problems	**Access to medical care**	• Long-time (4 years) between flood and data collection • Potential recall bias and underreporting of health conditions • Questions lack validity and accuracy • Unknown range of ages for participants
					Adverse event experiences	Access to medical care	

(continued)

Table 4.2 (continued)

Reference(s)	Design, country, and period of data collection	Disaster	Age (% or mean ± SD)/gender (%)	Sample size of analysis	Measures of NRR outcomes	Measures of elements Bold if significant association with NRR outcome	Limitations
Gordon et al. [25]	Cross-sectional study, USA, 2009	Red River Flood of 2009	19.77 ± 2.59 Men: 50% Female: 50%	210	Acquired Capability for Suicide Scale	Number of hours spent volunteering	• May not have sufficient power in sample size • Unclear if confounders like age and gender were controlled
					Perceived thwarted belongingness	**Number of hours spent volunteering**	
					Perceived burdensomeness	**Number of hours spent volunteering**	
Huang et al. [29]	Cross-sectional study, China, 2000	Floods of 1998	16–34: 44.7% 35–54: 39.5% 55+: 15.8% Male: 47.1% Female: 52.9%	17,846	Probable PTSD	**Perceived mental status before and after flood**	• Potential recall bias • Potential Information bias • Potential misreporting of PTSD
Lamond et al. [36]	Cross-sectional study, United Kingdom, 2013	Summer floods of 2007	Not provided	280	Severity of deterioration of mental health	**Household income level** **Registered for future flood warning** Received flood warning **Action based on flood warning** **Moved high value items** Relocated the kitchen **Implemented sandbags** **Implemented window and door guards** Fitted airbrick and vent covers Waterproofed external walls Installed non-return valves Implemented automatic airbrick and/or vent **Need for relocation** **Length of relocation**	• May not have sufficient power in sample size • Long-time (6 years) between flooding and data collection • Potential recall bias and information bias • Unknown accuracy and validity on measures used
					Reinstatement costs due to flood damage	Received flood warning Action based on flood warning	

(continued)

Table 4.2 (continued)

Reference(s)	Design, country, and period of data collection	Disaster	Age (% or mean ± SD)/gender (%)	Sample size of analysis	Measures of NRR outcomes	Measures of elements Bold if significant association with NRR outcome	Limitations
Mason et al. [43]	Cross-sectional study, United Kingdom, 6 months after floods	Floods of 2007	$n = 444$ 56.87 ± 14.87 Male: 41.1% Female: 58.9%	392	PSTD (HTQ-R)	Physical health status **Evacuated the property** Rational coping Detached coping **Emotional coping** **Avoidance coping**	• No reasoning given for discrepancy between total sample size and sample used for analysis • Potential over-reporting of outcomes as assessed symptoms associated and not criteria that leads to diagnosis • Lack of specificity on how long mailed responses were given to return and be included in study
					Anxiety (HSCL)	**Physical health status** **Evacuated the property** **Rational coping** **Detached coping** **Emotional coping** Avoidance coping	
Mason et al. [43] [cont.]				393	Depression (HSCL)	**Physical health status** **Evacuated the property** Rational coping Detached coping **Emotional coping** Avoidance coping	
Norris et al. [48]	Longitudinal prospective study, Mexico, 2000	Floods of 1999	Mean age: 37 Male: 32% Female: 68%	658	Social embeddedness after disaster	Pre-disaster major depressive disorder	• Unequal proportion of gender • Possible language bias • Possible recall bias
					Household embeddedness	Pre-disaster major depressive disorder	
					Family embeddedness	Pre-disaster major depressive disorder	
					Friend embeddedness	Pre-disaster major depressive disorder	

(continued)

Table 4.2 (continued)

Reference(s)	Design, country, and period of data collection	Disaster	Age (% or mean ± SD)/gender (%)	Sample size of analysis	Measures of NRR outcomes	Measures of elements Bold if significant association with NRR outcome	Limitations
Phifer and Norris [59], Norris and Murrell [50]	Longitudinal prospective study, USA, 1981–1985	Floods in 1981 and 1984	55–59: 19.1% 60–64: 25.8% 65–74: 36.4% 75+: 18.7% Male: 38.6% Female: 61.4%	203	Trait Anxiety (STAI-T) (measured after 1984 flood)	**Trait Anxiety before the 1981 flood** Prior trauma	• Small sample size with unequal % of gender and age groups • Poorly explained study, unclear descriptions on methodology, measured variables, data collection, and analysis • Unclear what they intend to do with findings, especially using STAI-T from two floods • Lengthy time between flood and data collection (i.e., measured over a year after 1984 flood in 1985) • Potential recall bias
					Personal loss	**Weather-specific distress**	
					Community destruction	**Weather-specific distress**	
Norris et al. [54]	Cross-sectional study, USA, 2005–2006	Hurricane Katrina 2005	42 ± 5 Male: 68% Female: 32%	80	Number of PTSD symptoms (CIDI-PTSD)	Pre-Katrina mental health Time in transition camp	• Small sample size • Possible discrepancy of understanding and responding to questions due to language bias • Unclear on validity of measures
					Property damage	Pre-Katrina mental health Time in transition camp	
Paranjothy et al. [55], Greene et al. [26]	Cross-sectional study, UK, 2007–2008	Summer Floods of 2007	South Yorkshire 50 ± 17 Male: 28% Female: 72% Worcestershire 57 ± 17 Male: 43% Female: 57%	2,265	Psychological distress	**Self-reported health concerns** **Prior existing medical problems** Self-perceived impact on house value Requests for evacuations **Compliance to requested evacuations**	• Different proportion of gender, age, employment status, and risk levels of mental health symptoms between sample cities • Different collection methods for each city
					Probable anxiety	**Self-reported health concerns** **Prior existing medical problems** Self-perceived impact on house value Requests for evacuations Compliance to requested evacuations	

(continued)

Table 4.2 (continued)

Reference(s)	Design, country, and period of data collection	Disaster	Age (% or mean ± SD)/gender (%)	Sample size of analysis	Measures of NRR outcomes	Measures of elements Bold if significant association with NRR outcome	Limitations
Greene et al. [26], Paranjothy et al. [55] [cont.]					Probable depression	**Self-reported health concerns** **Prior existing medical problems** Self-perceived impact on house value Requests for evacuations Compliance to requested evacuations	
					Probable PTSD	**Self-reported health concerns** **Prior existing medical problems** Self-perceived impact on house value Requests for evacuations Compliance to requested evacuations	
				2,029	Psychological distress (GHQ-12)	**Social support and cohesion**	
Stimpson et al. [72]	Prospective study, USA, 1992–1993	Summer floods of 1993	52.1 ± 15.8 Male: 36% Female: 64%	1,735	Sought flood-related help after the flood	**Social support** **Previous flood experience**	• Most participants had very few flood experience and exposure • Unclear about drop-outs during follow-up • Other types of services, such as flood insurance, could had influenced the outcome

(continued)

Table 4.2 (continued)

Reference(s)	Design, country, and period of data collection	Disaster	Age (% or mean ± SD)/gender (%)	Sample size of analysis	Measures of NRR outcomes	Measures of elements Bold if significant association with NRR outcome	Limitations
Tunstall et al. [74]	Cross-sectional study, UK, 2002–2003	Floods since 1998	Based on 813 flooded people: 18–34: 10% 35–54: 42% 55–74: 36% 75+: 12%	507	Psychological distress (GHQ -12) at worst time	**Prior health status** **Evacuation** **Warning time** **Problems with insurers** **Time to get back to normal** **Rented accommodation**	• Varying number of sample sizes used for analysis • Unknown proportion of age, gender and other potential confounders for sample size used for analysis • Potential recall bias and information bias • Sample included mixture of people who been flooded in different years and unclear when • Limited descriptions on variables • Unclear if confounders were controlled
			Male: 39% Female: 61%	733	Psychological distress (GHQ-12) at current time	**Prior health status** **Problems with insurers** **Area house prices** **Time to get back to normal** Help received	
				629	PTSD Intensity Scores	**Prior Health status** **Evacuation** **Warning time** **Problems with insurers** **Time to get back to normal**	

Table 4.3 Evidence table for the outcome of psychological impact

Reference(s)	Measures of elements	Identified element	Reported association(s) with the psychological impact
Alderman et al. [4]	Existing health problems	Health	If having pre-existing health conditions than not having: • Higher odds for increased psychological distress (*aOR* = 4.4, 95% CI [2.7, 7.4]) • Higher odds of having probable PTSD (aOR = 2.9, 95% CI [1.5, 5.5]) • Problems with sleeping (aOR = 3.3, 95% CI [1.8, 6.2] • Worse overall health since the floods (aOR = 2.9, 95% CI [1.5, 5.7])
	Owning their property	Economic investment	If renting a property than owning: • Higher odds for problems with sleeping (*aOR* = 2.6, 95% CI [1.4, 4.6]) • Higher odds for probable PTSD (*aOR* = 2.6, 95% CI [1.4, 4.7])
Bei et al. [5]	Level of social support received	Community networks and relationships	Between the groups with no support, personal support only, government/community support only, and personal and government/community support: • Significant differences in post-flood PTSD symptoms (IES-R, $\chi^2 = 10.57, p < 0.01$) Between groups who did and did not evacuate: • Significant differences in scores of post-flood PTSD symptoms (IES-R, $U = 151.50, Z = -2.99, p < 0.01$)
	Evacuation status	Governance and leadership	• No significant association between higher depressive symptoms and evacuation ($F(1,45) = 3.73, p = 0.06, \eta^2 = 0.08$)
Bokszczanin [7, 8]	Availability of social support from parents	Community networks and relationships	• Found to be negatively correlated to higher incidence of PTSD symptoms ($r = -0.28, p < 0.001$) • After controlling for gender, age, place of residence, and exposure to trauma, "lower levels of parental support [$\beta = -0.11, p < 0.01$], higher levels of family conflict [$\beta = 0.21, p < 0.001$], and higher levels of overprotectiveness [$\beta = 0.28, p < 0.001$] predicted greater PTSD symptomatology" [8], p. 330)
Chen et al. [13]	Previous traumatic experiences	Health	• No significant associations between psychological distress and the mental health subscale of the SF36
Collins et al. [14]	Access to medical care	Health	• Found that "access to medical care is associated with a 70% decrease in the likelihood of post-flood mental health problems" [14], p. 421) ($aOR = 0.30, p < 0.01$)

(continued)

Table 4.3 (continued)

Reference(s)	Measures of elements	Identified element	Reported association(s) with the psychological impact
Gordon et al. [25]	Number of hours spent volunteering	Community networks and relationships	• No significant association between levels of acquired capability for suicide and number of volunteering hours ($\beta = 0.12$, $t = 1.82$, $p = 0.07$)
Huang et al. [29]	Perceived mental status before and after flood	Mental outlook	In a logistic regression model that included age, gender, education, type of flood, severity of flood, flood experience, and mental status: • Higher odds for probable PTSD were found among participants who scored higher for worse mental status before the flood ($aOR = 1.68$, 95% CI [1.601, 1.763]) and after the flood ($aOR = 1.957$, 95% CI [1.807, 2.120])
Lamond et al. [36]	Household income level	Economic investment	• Found to be negatively correlated with the severity of deterioration of mental health ($r = -0.17$, $p < 0.01$) In reference to household income level above £55,000: • Higher odds for severe mental health deterioration were found for participants with household income level less than £5,000 ($OR = 8.06$, 95% CI [1.4, 47.0])
	Registered for future flood warning	Preparedness	• Significant correlation with severity of mental health deterioration ($r = 0.148$, $p < 0.05$)
	Received flood warning	Communication	• No significant correlation with severity of mental health deterioration
	Action based on flood warning	Preparedness	• Found negative correlation with the severity of deterioration of mental health ($r = -0.14$, $p < 0.05$)
	Moved high value items	Preparedness	• Significant correlation with severity of mental health deterioration ($r = 0.278$, $p < 0.01$) • Lower odds for severe mental health deterioration if moving high value items to the first floor after the flood ($OR = 0.21$, 95% CI [0.1, 0.5])
	Relocated the kitchen	Preparedness	• No significant correlation with severity of mental health deterioration

(continued)

Table 4.3 (continued)

Reference(s)	Measures of elements	Identified element	Reported association(s) with the psychological impact
	Implemented sandbags	Preparedness	• Significant correlation with severity of mental health deterioration ($r = 0.204$, $p < 0.01$)
	Implemented window and door guards	Preparedness	• Significant correlation with severity of mental health deterioration ($r = 0.146$, $p < 0.05$) • Lower odds for severe mental health deterioration if installing door and window guards in their home after the flood ($OR = 0.28$, 95% CI [0.1,0.8])
	Fitted airbrick and vent covers	Preparedness	• No significant correlation with severity of mental health deterioration
	Waterproofed external walls	Preparedness	• No significant correlation with severity of mental health deterioration
	Installed non-return valves	Preparedness	• No significant correlation with severity of mental health deterioration
	Implemented automatic airbrick and/or vent	Preparedness	• No significant correlation with severity of mental health deterioration
	Need for relocation	Governance and leadership	• Significant correlation with severity of mental health deterioration ($r = 0.263$, $p < 0.01$)
	Length of relocation	Governance and leadership	• Significant correlation with severity of mental health deterioration ($r = 0.166$, $p < 0.05$) In reference to not relocating: • Higher odds for severe mental health deterioration were found for participants with duration of relocation longer than six months ($OR = 5.80$, 95% CI [2.2, 15.1])
Mason et al. [43]	Physical health status	Health	• "Participants who reported being in poor health were between 2 and 4.2 times more likely to be meet criteria for PTSD, anxiety and depression than participants who reported being in good health" ([43], p. 70) • Lower odds for anxiety were found for participants with good health status ($aOR =$ 0.236, 95% CI [0.116, 0.484]) • Lower odds for depression were found for participants with good health status ($aOR =$ 0.36, 95% CI [0.17, 0.75])

(continued)

Table 4.3 (continued)

Reference(s)	Measures of elements	Identified element	Reported association(s) with the psychological impact
	Evacuated the property	Governance and leadership	• Higher odds for PTSD were found for participants who vacated their home ($aOR = 2.04$, 95% CI [1.11, 3.75]) • Higher odds for anxiety were found for participants who vacated their home ($aOR = 2.18$, 95% CI [1.13–4.20]) • Higher odds for depression were found for participants who vacated their home ($aOR = 2.30$, 95% CI [1.27, 4.18])
	Rational coping	Mental outlook	• Higher odds for anxiety were found for participants who scored high for rational coping ($aOR = 2.74$, 95% CI [1.17, 6.40])
	Detached coping	Mental outlook	• Lower odds for anxiety were found for participants who scored high for detached coping ($aOR = 0.34$, 95% CI [0.13, 0.91])
	Emotional coping	Mental outlook	• Higher odds for PTSD were found for participants who scored high for emotional coping scale ($aOR = 8.86$, 95% CI [3.71, 21.18]) • Higher odds for anxiety were found for participants who scored high for emotional coping scale ($aOR = 8.87$, 95% CI [3.56. 22.09]) • Higher odds for depression were found for participants who scored high for emotional coping scale ($aOR = 23.66$, 95% CI [8.94, 62.60])
	Avoidance coping	Mental outlook	• Higher odds for PTSD were found for participants who scored high for avoidance coping scale ($aOR = 2.86$, 95% CI [1.29, 6.34])
Norris and Murrell [50], Phifer and Norris [59]	Prior (non-disaster) trauma	Health	• Prior trauma was not a significant predictor of participants having trait anxiety after the 1984 flood
	Trait Anxiety before the 1981 flood	Health	• Trait anxiety measured before the 1981 flood was found to predict trait anxiety measured in 1985 after the 1984 flood ($R^2 = 0.210, p < 0.001$) • Trait anxiety before the 1981 flood to predict the trait anxiety measured after the 1984 flood ($R^2 = 0.188, p < 0.001$) Among participants with prior flood exposure or with prior trauma experience: • Trait anxiety measured before the 1981 flood was found to predict trait anxiety measured after the 1984 flood (respectively, $R^2 = 0.437, p < 0.001$; $R^2 = 0.246, p < 0.001$)

(continued)

Table 4.3 (continued)

Reference(s)	Measures of elements	Identified element	Reported association(s) with the psychological impact
Norris et al. [54]	Pre-Katrina mental health	Health	• No significant association with Katrina-related PTSD symptoms
	Time in transition camp	Governance and leadership	• No significant association with Katrina-related PTSD symptoms
Greene et al. [26], Paranjothy et al. [55]	Self-reported health concerns after the flood	Mental outlook	• Higher odds of psychological distress for participants who reported higher perceived health concerns ($aOR = 4.7$, 95% CI [3.5, 6.4]) • Higher odds of probable anxiety for participants who reported higher health concerns ($aOR = 3.7$, 95% CI [2.6, 5.1]) • Higher odds of probable depression for participants who reported more health concerns ($aOR = 4.3$, 95% CI [3.0, 6.1]) • Higher odds of probable PTSD for participants who reported more health concerns ($aOR = 3.0$, 95% CI [1.9, 4.7])
	Prior existing medical problems	Health	• Higher odds of psychological distress for participants who reported existing medical problems ($aOR = 2.4$, 95% CI [1.8, 3.2]) • Higher odds of probable anxiety for participants who reported existing medical problems ($aOR = 2.4$, 95% CI [1.7, 3.5]) • Higher odds of probable depression for participants who reported existing medical problems ($aOR = 2.9$, 95% CI [2.0, 4.2]) • Higher odds of probable PTSD for participants who reported existing medical problems ($aOR = 4.1$, 95% CI [2.6, 6.7])
	Self-perceived impact on house value	Economic investment	• No association was found with the measures on psychological impact
	Requests for evacuations but refused	Communication	• No association was found with the measures on psychological impact

(continued)

Table 4.3 (continued)

Reference(s)	Measures of elements	Identified element	Reported association(s) with the psychological impact
	Compliance to requested evacuations	Preparedness	• Higher odds of psychological distress for participants who were asked to evacuate and who evacuated ($aOR = 1.7$, 95% CI [1.2, 2.5])
	Social support and cohesion	Community networks and relationships	• Negatively associated with psychological distress ($\beta = -0.162$, 95% CI [−0.244, −0.079])
Tunstall et al. [74]	Prior health status	Health	Using multivariate regression: • Significant association found with GHQ-12 scores at the worst time and poor prior health ($\beta = -0.14, p < 0.001$) • Significant association found with current GHQ-12 scores and poor prior health ($\beta = -0.14, p < 0.001$) • Significant association found with high PTSD intensity scores and poor prior health ($\beta = -0.19, p < 0.001$)
	Being evacuated	Governance and leadership	Using multivariate regression: • Significant association found with GHQ-12 scores at the worst time ($\beta = 0.13, p < 0.01$) • Significant association found with high PTSD intensity scores ($\beta = 0.12, p < 0.01$)
	Greater warning time before a flood occurred	Communication	Using multivariate regression: • Significant association found with GHQ-12 scores at the worst time ($\beta = -0.09, p < 0.05$) • Significant association found with high PTSD intensity scores ($\beta = -0.10, p < 0.01$)
	Problems with insurers	Economic investment	Using multivariate regression: • Significant association found with GHQ-12 scores at the worst time ($\beta = 0.27, p < 0.001$) • Significant association found with current GHQ-12 scores ($\beta = 0.07, p < 0.05$) • Significant association found with high PTSD intensity scores and participants ($\beta = 0.26, p < 0.001$)

(continued)

Table 4.3 (continued)

Reference(s)	Measures of elements	Identified element	Reported association(s) with the psychological impact
Tunstall et al. [74] [cont.]	Taking long time to get back to normal	Governance and leadership	Using multivariate regression: • Significant association found with GHQ-12 scores at the worst time ($\beta = 0.12, p < 0.01$) • Significant association found with current GHQ-12 scores ($\beta = 0.08, p < 0.01$) • Significant association found with high PTSD intensity scores ($\beta = 0.09, p < 0.05$)
	Prior accommodation type of renting	Economic investment	• Using multivariate regression, significant association found with GHQ-12 scores at the worst time ($\beta = 0.09$, $p < 0.05$)
	Amount of help received	Community networks and relationships	• No significant association was found with current GHQ-12 scores
	Area house prices	Economic investment	• Using multivariate regression, significant association found with current GHQ-12 scores ($\beta = -0.08$, $p < 0.05$)

4.3.2.1 Alderman et al. [4]

Alderman et al. [4] carried out a cross-sectional survey with 960 respondents from neighborhoods in Australia that suffered direct and indirect impacts of a flood in 2011. Although an adequately powered sample size, they estimated that only about 12% of their respondents reported being directly affected by flooding. The authors measured the participant's amount of psychological distress using a short version of Kessler 10 & Kessler 6 [psychological impact] [31], the levels of PTSD using the Posttraumatic Stress Disorder-Civilian Checklist (PCL-C scale) [psychological impact] [80], and the sleep quality using the Groninger Sleep Quality Scale (GSQS-10) [psychological impact] [42] caused by the flood. These were tested against the participant's existing health problems [health] and whether they owned their own property [economic investment]. Participants had higher odds of increased psychological distress if they had pre-existing health conditions ($aOR = 4.4$, 95% CI [2.7, 7.4]). Additionally, those with existing health problems had higher odds of having probable PTSD ($aOR = 2.9$, 95% CI [1.5, 5.5]), problems with sleeping ($aOR = 3.3$, 95% CI [1.8, 6.2], and worse overall health since the floods ($aOR = 2.9$, 95% CI [1.5, 5.7]) than those who had no existing health problems. People who were renting their property were found to have higher odds of problems with sleeping ($aOR = 2.6$, 95% CI [1.4, 4.6]) and probable PTSD ($aOR = 2.6$, 95% CI [1.4, 4.7]) than those who owned their property. Although associations between psychological impact and the elements of *health and economic investment* were significant, a slight discrepancy was found between the numbers of respondents for this study's survey, ranging from 920 to 960 people depending on the questionnaire. An explanation would help clarify whether this missing data was due to dropouts or other issues. Also, most of the participants were older than 45 years old, limiting the results to a particular age group.

4.3.2.2 Bei et al. [5]

In a longitudinal prospective study, Bei et al. [5] received questionnaire responses from 274 participants in Australia who completed them before and after a flood between the years 2010 and 2011. However, the statistical association tests only used a smaller portion of this total. The questionnaire included measures on flood experience, the impact of event scales (IES-R) [psychological impact] [82], center for epidemiological studies depression scale (CES-D) [psychological impact] [28], and medical outcomes short form 12 (SF-12) [psychological impact] [77]. These outcomes were assessed with the participant's level of social support received [community networks and relationships] and the participant's evacuation status [governance and leadership]. Evacuation status was seen as part of an element because an assumption was made that governance or leadership had to be taken to initiate and safely evacuate. The authors did not find a significant association between higher depressive symptoms and evacuation ($F(1,45) = 3.73$, $p = 0.06$, η^2

= 0.08). However, based on 45 responses, there were significant differences in post-flood PTSD symptoms (IES-R, $\chi^2 = 10.57$, $p < 0.01$) and post-flood perceived physical health ($SF12_{physical}$, $\chi^2 = 7.90$, $p < 0.05$) between the groups with no support, personal support only, government/community support only, and personal and government/community support. Based on 53 responses, there were significant differences in scores of post-flood PTSD symptoms between those who did and did not evacuate (IES-R, $U = 151.50$, $Z = -2.99$, $p < 0.01$). Bei et al. ([5], p. 992) concluded that "greater flood exposure and the lack of social support were the risk factors for poorer mental and physical health." Although these associations were significant, it was unclear of the timing given to answer the survey after the flood. Despite timing and small sample size, evidence found suggested that the elements of community networks and relationships and governance and leadership made a difference in the psychological impact seen in flood-affected participants.

4.3.2.3 Bokszczanin [7, 8]

Bokszczanin [7] carried out a study 28 months after a flood in southwestern Poland in 1997. The goal was to identify the predictors of PTSD in 522 adolescents who were aged 11 to 21 years old. In a later publication, Bokszczanin [8] used this same sample and carried out additional calculations with the amount of PTSD symptoms measured by using the Revised Civilian Mississippi PTSD scale [psychological impact] [51]. In this later publication, the levels of parental support among the children and adolescents [community networks and relationships] were found to be negatively correlated to a higher incidence of PTSD symptoms ($r = -0.28$, $p < 0.001$). After controlling for gender, age, place of residence, and exposure to trauma, "lower levels of parental support [$\beta = -0.11, p < 0.01$], higher levels of family conflict [$\beta = 0.21, p < 0.001$], and higher levels of overprotectiveness [$\beta = 0.28, p < 0.001$] predicted greater PTSD symptomatology" ([8], p. 330).

This study's quality was determined low in both publications due to the unclear and brief information provided about its "correlational and retrospective design" ([7], p. 350), which was described more as a retrospective cross-sectional survey. Recall bias could also occur as the study was implemented 28 months after a flood. Also, the study was unclear on how the sample size of analysis was determined and the composition of gender and age for this sample. Despite this, the study found higher parental support to reduce the likely PTSD symptoms in adolescents affected by a flood.

4.3.2.4 Chen et al. [13]

Using self-reported survey data collected between 2006 to 2007, Chen et al. [13] examined 113 Vietnamese-American survivors of Hurricane Katrina on measures of psychological distress [psychological impact], measured by IES-R [82] and health-related quality of life (which included mental and physical health), measured by

SF-36 [78] [psychological impact]. Non-significant associations were found for psychological distress and the mental health subscale of the SF36, as it relates to the participant's previous traumatic experiences before Hurricane Katrina [health]. However, after controlling for acculturation, time since Hurricane Katrina, age, and gender, participants with a more significant amount of previous traumatic experiences were also associated with reporting poorer physical health, ($\beta = -0.29$, $p < 0.01$).

This study may not have sufficient power in sample size and may include a language barrier or bias during collection. Not all the participants may have been given and/or responded in the same language. Possible recall bias could have occurred as longer than a year had passed between the flood and data collection. Though this study did not demonstrate an association between psychological impact and the community resilience element of health, an association was found between prior traumatic experiences and poorer physical health among flood-affected participants.

4.3.2.5 Collins et al. [14]

Collins et al. [14] carried out a survey in 2010 with 475 individuals from households that were affected by the 2006 floods in Texas, USA. They asked participants about their mental health problems experienced within four months of the 2006 floods [psychological impact] and their access to health services [health]. Determining access to health services was based on one question asked, "do you have a regular doctor or a clinic where you can go for your routine medical care" ([14], p. 418). The authors found that "access to medical care is associated with a 70% decrease in the likelihood of post-flood mental health problems" ([14], p. 421) ($aOR = 0.30$, $p < 0.01$). Although this association was found to be significant, it must be noted that recall bias could have occurred as the data collection took place four years after the floods. Participants were asked to recall the experience within four months after the floods in the data collection. This could also contribute to possible underreporting of health conditions, as some may have been forgotten soon after the floods. Another limitation seen was that determining access to medical care was based on a question about if a participant has a regular doctor or clinic and did not consider whether a person could have the means to travel to see a regular doctor or clinic. However, a strength of this study was that the authors did their best to track down flooded households from 2006 at the time of data collection in 2010, even though it was unclear if all homes in the survey reported flood damage in 2006. This study showed that access to health services mattered to post-flood mental health impact.

4.3.2.6 Gordon et al. [25]

Gordon et al. [25] carried out a study on the interpersonal risk factors for suicide among a sample size of 210 undergraduate students who were affected by the 2009 Red River Flood in North Dakota, USA. They included the Acquired Capability for

Suicide Scale [76] [psychological impact] and asked about the number of volunteer hours that participants did during the flood [community networks and relationships]. The Acquired Capability for Suicide Scale measured the participant's fearlessness about committing suicide, although this study did not capture information on the participant's suicide behaviors. Even though this study was unclear whether the analysis was controlled for confounders, such as age and gender, the authors may not have sufficient power in their sample size. For example, they did not find a significant association between levels of acquired capability for suicide and the number of volunteering hours ($\beta = 0.12$, $t = 1.82$, $p = 0.07$). A larger sample size could be more telling.

4.3.2.7 Huang et al. [29]

In a cross-sectional study by Huang et al. [29], 25,478 people affected by the 1998 floods in China were interviewed in 2000. The interviews used a questionnaire "to obtain demographic data, to ascertain PTSD, and to measure personality and psychological characteristics" ([29], p. 2). The measures of perceived mental outlook status were assessed before and after a flood [mental outlook]. Before the flood, the authors created this measure by including questions that asked participants if they considered themselves tense or highly strung, felt that life was boring or lonely and if they were easily hurt when people found faults with them or their work. After the flood, the authors created a variable based on responses to the following questions: whether participants were trapped and waited for rescue during the flood, whether they or relatives were seriously injured during the flood, whether they witnessed others drowning, whether their home was damaged, and whether this flood was their first experience of floods. These variables were measured among 17,846 people, randomly selected as a subgroup from the affected population for association with probable PTSD scores [psychological impact]. In a logistic regression model that included age, gender, education, the type of a flood, the severity of a flood, flood experience, and mental status, higher odds for probable PTSD were found among participants who scored higher for worse mental status before the flood ($aOR = 1.68$, 95% CI [1.601, 1.763]) and after ($aOR = 1.957$, 95% CI $[1.807, 2.120]$). This study may have a recall and information bias since the survey occurred two years after the flood. Also, the diagnosis of PTSD in this study may not be accurate as the diagnosis was made by interviewers who were not trained as psychologists. Although psychologists supervised the interviewers, the study was unclear as to the level of supervision by the psychologists. Still, this study had an adequately powered sample size, and the association between mental outlook and the psychological impact was found to be significant.

4.3.2.8 Lamond et al. [36]

In a cross-sectional study, Lamond et al. [36] mailed surveys to households flooded in the 2007 floods in England. The flooded household sample was "not designed to be representative of national socio-demographic pattern" ([36], p. 328). According to a previously published paper, "this survey was carried out in 2013 some 6 years after the event" ([30], p. 179). Mailing locations were selected from areas with greater than 50 affected properties and from which participants had submitted insurance claims. Two hundred eighty responses from 2,309 mailed questionnaires were included in this study's results. Lamond et al. [36] correlated severe mental health deterioration [psychological impact] with the household income level and whether the participants prevented flood damages [preparedness], received a flood warning [communication] and compliance before a flood [preparedness], and the need for and length of relocation [governance and leadership]. Severe mental health deterioration "was gathered using a five-point Likert scale ranging from 'extreme impact' to 'no impact,'" and participants "were required to assess and report the deterioration in psychological health for their household as a whole" ([36], p. 327). Actions to prevent further flood damages included registering for future flood warnings, moving high-value items, relocating the kitchen, implementing sandbags, implementing window and door guards, fitting airbrick and vent covers, waterproofing external walls, installing non-return valves, and implementing automatic airbrick or vent. The need for relocation and the length of relocation was seen as the element of governance and leadership because they may reflect the initiation of governance or leadership for post-flood infrastructure and services.

Household income level was found to be negatively correlated with the severity of deterioration of mental health ($r = -0.17, p < 0.01$). Also, a negative correlation with the severity of deterioration of mental health and action as a result of the flood warning to prevent damage was found ($r = -0.14$, $p < 0.05$). Severity of mental health deterioration was also significantly correlated with: the need for relocation ($r = 0.263, p < 0.01$); length of relocation time ($r = 0.166, p < 0.05$); being registered for future flood warnings ($r = 0.148, p < 0.05$); use of sandbags, ($r = 0.204, p < 0.01$); moving high value items to the first floor ($r = 0.278$, $p < 0.01$); and use of doors and window guards ($r = 0.146$, $p < 0.05$). Relocating the kitchen to the first floor, fitting airbrick and vent covers, waterproofing external walls, installing non-return valves, and implementing automatic air-brick and/or vent did not correlate with mental health impact.

Examining severe mental health deterioration against pre-, during, and post-flooding factors, Lamond et al. [36] found only two factors with higher odds for severe mental health deterioration: household income level and duration of the relocation. In reference to household income levels above £55,000, higher odds for severe mental health deterioration were found for participants with household income levels less than £5,000 ($OR = 8.06$, 95% CI [1.4, 47.0]), which is far below the UK poverty line. In reference to not relocating, higher odds for severe mental health deterioration were found for participants with a duration of relocation longer than six months ($OR = 5.80$, 95% CI [2.2, 15.1]). Two factors were found among the participants to have lower odds of severe mental health deterioration: moving high-value items to the first floor after the flood (*OR* = 0.21, 95% CI [0.1, 0.5]) and installing door and window guards in their home after the flood (*OR* = 0.28, 95% CI [0.1, 0.8]). Although associations were found significant, it must be noted that this study included a small and non-representative sample of the UK, the length of time between the flood and data collection was long (6 years). Like other studies identified in this review, the study was cross-sectional, so causality cannot be inferred. Additionally, many questions were self-developed and unknown in their validity and accuracy as a measurement. Still, this study found significant associations between psychological impact and the elements of *economic investment, preparedness, and governance and leadership.*

4.3.2.9 Mason et al. [43]

In another cross-sectional survey looking at the floods of 2007 in the United Kingdom, Mason et al. [43] mailed household questionnaires to two local council areas six months after the floods. Still, it was unknown for how long they accepted responses. The questionnaire included demographics, physical health status [health], whether evacuation occurred [governance and leadership], flood experience questions, type of community area, the Harvard Trauma Questionnaire (HTQ-R) [psychological impact] [45], the Hopkins Symptom Checklist (HSCL) [psychological impact] [56], and the Coping Styles Questionnaire (CSQ) [mental outlook] [66]. The HTQ-R measured PTSD, while the HSCL measured anxiety and depression. CSQ assessed the participants' coping strategies based on the adaptive coping styles of rational and detached coping and the maladaptive coping techniques of emotional and avoidance coping.

Mason et al. ([43], p. 70) found that "participants who reported being in poor health were between 2 and 4.2 times more likely to be meet criteria for PTSD, anxiety, and depression than participants who reported being in good health." Additionally, higher odds for PTSD were found for participants who vacated their home (*aOR* = 2.04, 95% CI [1.11, 3.75]), participants who scored high for emotional coping scale (*aOR* = 8.86, 95% CI [3.71, 21.18]), and participants who scored high for avoidance coping scale (*aOR* = 2.86, 95% CI [1.29, 6.34]). Higher odds for anxiety were found for participants who vacated their home (*aOR* = 2.18, 95% CI [1.13–4.20]), and participants who scored high for rational coping (*aOR* = 2.74, 95% CI [1.17, 6.40]), and emotional coping scale (*aOR* = 8.87, 95% CI [3.56. 22.09]). Additionally, higher odds for depression were found for participants who vacated their home (*aOR* = 2.30, 95% CI [1.27, 4.18]) and participants who scored high for emotional coping scale (*aOR* = 23.66, 95% CI [8.94, 62.60]). Lower odds for anxiety were found for participants with good health status (*aOR* = 0.236, 95% CI [0.116, 0.484]) and participants who scored high for detached coping (*aOR* = 0.34, 95% CI [0.13, 0.91]). Lower odds for depression were found for participants with good health status (*aOR* = 0.36, 95% CI [0.17, 0.75]).

Although Mason et al. [43] found significant associations between psychological impact and the elements of health, governance and leadership, and mental outlook, it must be noted that no reasoning was given for the discrepancy between the total sample size and sample used for analysis. It was also unclear about the data collection period (i.e., how long did the participants have to return the mailed survey). Potential recall bias and over-reporting of outcomes may have occurred. The outcomes were assessed by the associated symptoms and not by the clinical criteria that led to a diagnosis for the outcome.

4.3.2.10 Norris and Murrell [50], Phifer and Norris [59]

Norris and colleagues [50, 59] used a prospective design with five follow-up intervals between 1981 to 1985. This study examined participants affected by the 1981 and 1984 floods in Kentucky, USA. The authors assessed if participants with a prior (non-disaster) trauma [health] or with trait anxiety, measured before and after the 1981 flood [health], would have trait anxiety after the 1985 flood (trait half of State-Trait Anxiety Inventory, STAI-T) [psychological impact] [69]. Trait anxiety measured before and after the 1981 flood was seen as an element of health instead of mental outlook because this measurement is a clinical mental health scale, STAI-T.

Based on 203 adults who fully completed the questionnaire, trait anxiety measured before the 1981 flood predicted trait anxiety measured in 1985 after the 1984 flood ($R^2 = 0.210, p < 0.001$), which makes sense as a personality trait tends to be stable. Prior trauma was not a significant predictor of participants' trait anxiety after the 1984 flood ($R^2 = 0.001$). The authors also carried out further regressions based on participant's prior exposure to trauma and a flood, but it was unclear whether the prior exposure to a flood was based on a new sample recruited in 1985. Among participants with no trauma exposure, Norris and Murrell [50] found trait anxiety before the 1981 flood to predict the trait anxiety measured after the 1984 flood ($R^2 = 0.188, p < 0.001$). Among participants with prior flood exposure or prior trauma experience, trait anxiety measured before the 1981 flood was found to predict trait anxiety measured after the 1984 flood (respectively, $R^2 = 0.437, p < 0.001$; $R^2 = 0.246, p < 0.001$). Although these associations were significant, it must be noted that this study had vague and conflicting descriptions of methodology and data collection between the two publications. Also, using anxiety as a trait before and after floods may not inform much about the association of the elements of community resilience in relation to psychological impact but more about the stability of a trait personality of anxiety.

4.3.2.11 Norris et al. [54]

A year after Hurricane Katrina hit New Orleans, Norris et al. [54] assessed 82 Vietnamese American adults for PTSD symptoms. Participants were found through a May 2005 register of Vietnamese American families that was part of a more extensive study on immigration that occurred 1 to 3 months before Hurricane Katrina. The participants decided on the language used in the interviews for the 2005 immigration study and the 2006 post-Katrina study by Norris et al. [54]. Most participants preferred the interviews to be conducted in Vietnamese (93%) than English (7%), which could have a positive or negative effect depending on the discrepancy between understanding and answering the questions. The authors measured Katrina-related PTSD symptoms [psychological impact] by using a modified version of Module K of Version 2.1 of the Composite International Diagnostic Interview (CIDI) [84]. To make this outcome into continuous measures of PTSD, the authors created variables "by counting the number of affirmative responses to instructive symptoms (Count B, range: 0–5), avoidance/numbing symptoms (Count C, range: 0–7), arousal symptoms (Count D, range: 0–5), and all symptoms (Count BCD, range: 0–17)" ([54], p. 95). This was assessed with a pre-Katrina mental health measure [health], which used the Mental Component Score of the Short Form 36 [78] and was measured at baseline before the disaster. Also, the length of time, in which participants were in a transition camp during and after the disaster [governance and leadership], was tested for association with Katrina-related PTSD symptoms.

With a small sample of 80 participants, the authors found non-significant associations for pre-Katrina mental health ($\beta = -0.20$) and time in transition camp ($\beta = 0.04$) with Katrina-related PTSD symptoms. This was based on a regression model, which included participant's background variables, objective exposure to Katrina, and subjective trauma exposure.

Hurricane Katrina included storms that led to flooding, but it was unclear whether the participants perceived the disaster as only the flood, storms, or all when answering the survey. Also, the validity of the self-developed questions by the authors was unknown, and the study included a small sample size that may not have sufficient power to detect an association. This study did not find a significant association between the element of health and psychological impact.

4.3.2.12 Greene et al. [26], Paranjothy et al. [55]

Additional associations between psychological impact and elements of community resilience were found in a study by Paranjothy et al. [55]. In this study, the authors examined survey data from 2,265 people affected by the 2007 floods in the United Kingdom. The survey examined the outcomes of psychological distress (GHQ-12) [psychological impact] [24], probable anxiety (Generalized Anxiety Disorder) [psychological impact] [70], probable depression (Patient Health Questionnaire) [psychological impact] [33], and probable PTSD (PTSD checklist) [psychological impact] [40]. These outcomes were tested for association with the following: participant's prior existing medical problems [health], participant's health concerns after the flood [mental outlook], participant's perception of the impact on house value [economic investment], and whether the participant received a request for evacuation [communication] and its compliance prior the flood [preparedness]. Health concerns after the flood were seen as the participant's post-flood attitude, thus being considered as a mental outlook. Perception of the house value was seen as an example of economic investment because it informed the participant's perceived impact on a previous investment made prior to the flood (buying a home).

Odds ratios were used in this study and were adjusted for age, previous medical problems, gender, employment, and area and data collection method. Still, it is essential to note that this study had different collection times between South Yorkshire and Worcestershire, along with the different risk levels of mental health symptoms found in each city. Events could have happened between the towns during the different collection times that could have created a bias among the participants. The various risk levels of mental health symptoms in each city could also mean that the towns may have a population with multiple awareness levels of mental health. There was also a different proportion of gender, age, and employment status between cities.

For the outcome of psychological distress, Paranjothy et al. [55] found four significant associations. There were higher odds for psychological distress among those who reported higher perceived health concerns ($aOR = 4.7$, 95% CI [3.5, 6.4]) and those who reported existing medical problems ($aOR = 2.4$, 95% CI [1.8, 3.2]). Additionally, higher odds for psychological distress were found among participants who were asked to evacuate and who evacuated than those who were not asked ($aOR = 1.7$, 95% CI [1.2, 2.5]).

Paranjothy et al. [55] found two significant associations for the outcome of probable anxiety. Higher odds for probable anxiety were found for those who reported higher health concerns ($aOR = 3.7$, 95% CI [2.6, 5.1]) and those who reported existing medical problems ($aOR = 2.4$, 95% CI [1.7, 3.5]).

For the outcome of probable depression, Paranjothy et al. [55] found two significant associations. Higher odds for probable depression were found in participants who reported more health concerns ($aOR = 4.3$, 95% CI [3.0, 6.1]) and in participants who reported existing medical problems ($aOR = 2.9$, 95% CI [2.0, 4.2]).

For the outcome of probable PTSD, Paranjothy et al. [55] found two significant associations. Higher odds for probable PTSD were found in participants who reported more health concerns ($aOR = 3.0$, 95% CI [1.9, 4.7]) and participants who reported existing medical problems ($aOR = 4.1$, 95% CI [2.6, 6.7]). No association was found between the measures on mental health and self-perceived impact on house values.

Using the same data as Paranjothy et al. [55], Greene et al. [26] carried out further statistical analysis and used only 2,029 people from the same dataset. Using a Bayesian version of a structural equation modeling approach, they explored "both direct and indirect effects and allowing the production of latent variables derived through the correlation structure of directly observed variables" ([26], p. 1793). These latent variables were measured for association with psychological distress assessed using the GHQ-12 [psychological impact] [24]. The authors defined one latent variable as 'vulnerability,' which did not completely match with one distinct element of community resilience or one unique NRR security outcome. Questions related to services disrupted were marked as social disruption, and a question on previous flood experience matched with mental outlook, while the remaining questions on flood risk were not labeled as an element or outcome. The other latent variable was called 'resilience' by the authors [community networks and relationships]. This consisted of questions related to perceived social cohesion, measured via the Neighborhood Cohesion Scale [10, 65] and asked whether participants had family nearby and the number of years they had been in the area. This latent variable was found to be negatively associated with psychological distress (B = −0.162, 95% CI [−0.244, −0.079]); however, it should be noted that this latent variable is a better reflection of social support and cohesion in the community than a community's resilience. Overall, this study found the elements of health, mental outlook, and community networks and relationships to be associated with psychological impact.

4.3.2.13 Tunstall et al. [74]

In a publication by Tunstall et al. [74], research was carried out in 30 locations in England and Wales as part of a larger project for a technical report to the UK Environment Agency on the appraisal of the human-related intangible impacts of flooding (Risk and Policy Analysts Ltd. et al. 64). Depending on the analysis, the sample size in Tunstall et al. [74] varied from 507 to 733 people. Regressions included outcomes on mental health status due to the flood, based on the Post-Traumatic Stress Scale [67] and psychological distress measured using the GHQ-12 [psychological impact] [24]. The authors asked the participants to score the GHQ-12 twice to capture the participant's flooding experience at the time of the most severe health impacts (GHQ-12 worst time) and the participant's experience over the past few weeks of the survey being administered (GHQ-12 current). These outcomes were tested against prior health status [health], the amount of warning time before a flood occurred [communication], the amount of help received [community networks and relationships], and the participant's evacuation status [governance and leadership]. The amount of time needed to get back to normal [governance and leadership] was also assessed. It reflected the quality of infrastructure and services provided by the post-flood leadership or government. Participant's prior accommodation type [economic investment], problems with insurers [economic investment], and house prices [economic investment] were also examined against the outcomes.

Tunstall et al. [74] found several significant associations using multivariate regression analysis. Significant associations were found between GHQ-12 scores at the worst time and having had problems with one's insurers ($\beta = 0.27, p < 0.001$), being evacuated ($\beta = 0.13$, $p < 0.01$), taking a longer time to get back to normal ($\beta = 0.12, p < 0.01$), and living in rented accommodation ($\beta = 0.09$, $p < 0.05$). Further associations with high GHQ-12 scores at the worst time were found with having poor prior health issues ($\beta = -0.14$, $p < 0.001$) and having a greater warning time before a flood ($\beta = -0.09, p < 0.05$). For high current GHQ-12 scores, significant associations were found with participants having a longer time to get back to normal ($\beta = 0.08, p < 0.01$), having had problems with insurers ($\beta = 0.07, p < 0.05$). having poor prior health issues ($\beta = -0.14, p < 0.001$), and having high house prices ($\beta = -0.08, p < 0.05$). Significant associations were found between high PTSD intensity scores and participants who had problems with insurers ($\beta = 0.26, p < 0.001$), were evacuated ($\beta = 0.12$, $p < 0.01$), and had a longer time to get back to normal ($\beta =$

0.09, $p < 0.05$). Participants who had high PTSD intensity scores were also associated with poor prior health issues ($\beta = -0.19, p < 0.001$) and a longer warning time before the flood occurred ($\beta = -0.10, p < 0.01$).

It must be noted that this study included limited descriptions for variables that matched as elements of community resilience, and most had to be inferred during the presentation of the results. This could lead to misinterpretation of the study's results; in one example, the author's meaning of "area house prices" was interpreted as house value, which may not be accurate to what was asked. Also, the sample size included a mixture of people who had been flooded in different years. As a result, the time elapsed between the flood and data collection may not be consistent among all participants. This study may have varying degrees of recall and information bias. Still, this study found the psychological impact to be associated with the elements of economic investment, preparedness, and governance and leadership.

4.3.3 *Outcome of Social Disruption*

Associations between social disruption and an element of community resilience were found in six studies. Three were cross-sectional, and three were carried out in the USA. Most studies (67%) were focused on community networks and relationships. Overall, *community networks and relationships* were found to reduce the *social disruption* caused by the floods. *Health* was also found to be important in lowering social disruption seen in the study populations. Also, people with previous flood experience were more likely to seek help during the disruption caused by the flood than those who did not have previous flood experience. Table 4.4 captures the evidence of the associations with social disruption found in each study.

4.3.3.1 Bei et al. [5]

In a longitudinal prospective study in Australia, Bei et al. [5] examined life satisfaction post-flood. The authors used the satisfaction with life scale (SWLS) [social disruption] [18] to test its association with levels of social support received during the flood experience [community networks and relationships]. There was a significant difference in the level of satisfaction with life among the groups of no support, personal support only, government/community support only, and personal and government/community support (SWLS, $\chi^2 = 7.82, p < 0.05$). The worst scores for satisfaction with life among the four groups were those who received government/community support, but this effect was non-significant ($U = 49.00, p = 0.16$). It must be noted that this study may not have sufficient power in the sample size. This study was unclear on the exact association between social disruption and the element of community networks and relationships; other than, the participants who received government or community support had lower scores for being satisfied with life.

Table 4.4 Evidence table for the outcome of social disruption

Reference(s)	Measures of elements	Identified element	Reported association(s) with social disruption
Bei et al. [5]	Social support received	Community networks and relationships	• Significant difference in the level of satisfaction with life among the groups of no support, personal support only, government/community support only, and personal and government/community support (SWLS, $\chi^2 = 7.82, p < 0.05$) • Worst scores for satisfaction with life after the flood were people who received government/community support but the effect on the scores was non-significant ($U = 49.00, p = 0.16$)
Chang [12]	Community cohesion	Community networks and relationships	• Participants felt their community cohesion was stronger after the flood ($(t(95) = 37.82, p < 0.001)$)
	Community affect	Community networks and relationships	• No significant association was found with high severity of disruption
	Community cognition	Community networks and relationships	• Negatively correlated with high severity of disruptions caused by the floods ($r = -0.22, p < 0.5$)
	Community behavior	Community networks and relationships	• No significant association was found with high severity of disruption
Collins et al. [14]	Access to health services	Health	• No correlation was found with adverse experiences caused by the flood
Gordon et al. [25]	Higher amount of volunteer hours	Community networks and relationships	• Found to have lower levels of thwarted belongingness ($\beta = -0.12, t = -2.80, p < 0.01$) and perceived burdensomeness ($\beta = -0.08, t = -2.42, p = 0.02$)
Norris et al. [48]	Pre-disaster major depressive disorder	Health	• No significant associations found with social embeddedness or with its subscales (friend embeddedness, family embeddedness, and household embeddedness)
Stimpson et al. [72]	Social support	Community networks and relationships	• Participants with higher amounts of social support had lower odds of seeking flood-related help afterwards than those with lower amount of social support ($aOR = 0.55, p = 0.011$)
	Past flood experience	Mental outlook	• Participants with more previous flood experience had higher odds of seeking flood-related help afterwards than those with less previous flood experience ($aOR = 1.58, p < 0.0001$)

4.3.3.2 Chang [12]

After the 2005 floods in Carlisle, United Kingdom, Chang [12] mailed questionnaires to potential respondents found through community centers and local leaders. Based on 96 questionnaires, the author examined the severity of flood disruption [social disruption] and its association with community cohesion [community networks and relationships], community affect [community networks and relationships] (measured by the Sense of Community Index [58], community cognition [community networks and relationships] (measured by the Social Life Feeling Scale [54], and community behavior [community networks and relationships] (measured by the Community Participation Scale [61]). A scale to measure the severity of disruptions included problems with gas and electricity supply and delays in damage repairs. Community cohesion was measured as self-perceived changes between pre- and post-flood periods. Sample items for community affect included "I think my community is a good place for me to live and I can recognize most of the people who live in my community" ([12], p. 294). Sample items provided for community cognition included "people in this community help each other out" and "social relations between different ethnic groups in this community have improved" ([12], p. 294). Community behavior included items on the likelihood that the participants would, for example, "get involved in any local clubs, organizations or schemes" and "contribute either money or time to a scheme to improve the community" ([12], p. 295).

Chang [12] found high severity of disruptions caused by the floods to be negatively correlated with community cognition ($r = -0.22$, $p < 0.5$). The author also found that participants felt their community cohesion was stronger after the flood ($t(95) = 37.82$, $p < 0.001$). Non-significant associations between severity of disruptions and community affect and behavior were found. It must be noted that the study did not specify the collection period, such as how long after the flood the questionnaire was sent out and for how long questionnaires for this study were accepted. This study also did not have a baseline (pre-flood) measure, so the observed change in cohesion may not be due to the flood. This study's sample size may not have sufficient power to detect associations and missed the input from community members who permanently moved out, whose response may change the results. Also, it was unknown what level of direct or indirect impact the participants experienced due to the flood. Still, this study found participants to feel stronger community cohesion after the flood, and the correlation between more significant social disruption and lower scores for community cognition.

4.3.3.3 Collins et al. [14]

A survey by Collins et al. [14] examined the adverse experiences caused by a flood [social disruption] by aggregating 13 items related to a negative flood experience, such as being separated from children, fleeing from home because water started to flood the house, losing, or having to abandon a pet and fearing for one's life. No correlation was found between this outcome and access to health services [health].

Still, this study may not be capturing the essence of access to health services since the question used only focused on whether participants had a regular doctor or clinic.

4.3.3.4 Gordon et al. [25]

Gordon et al. [25] surveyed participants using the Interpersonal Needs Questionnaire [social disruption] [76] after a flood in the USA. The Interpersonal Needs Questionnaire included subscales of perceived thwarted belongingness and perceived burdensomeness. Perceived thwarted belongingness measured how disconnected the participant felt from others, while perceived burdensomeness assessed how the participant felt like a burden on others. Participants who had a higher amount of volunteer hours [community networks and relationships] were found to have lower levels of thwarted belongingness ($\beta = -0.12, t = -2.80, p < 0.01$) and perceived burdensomeness ($\beta = -0.08, t = -2.42, p = 0.02$); however, it was unclear if confounding variables, such as age and gender were included in the regressions used. Still, this study showed that helping within the community after a flood would likely reduce the social disruption in the community.

4.3.3.5 Stimpson et al. [72]

In a prospective study by Stimpson et al. [72], cohort data were collected via telephone in two waves, before and after the summer floods of 1993 in Iowa, USA. Interviews before the floods, which occurred in 1992, included 2,406 people. Of these, 1,735 people were re-interviewed about 60 days after the floods in 1993. The authors asked the participants if they sought flood-related help after the flood [social disruption]. Although many things could influence why someone was seeking help, this was seen as an outcome of social disruption. Some disruption or interruption of the participant's normal daily life may have occurred that led the participants to seek help after the flood. Questions were also asked about social support [community networks and relationships], measured by a 10-item index on social support assessing the participant's perception of who was there for them and who could help them, and past flood experience [mental outlook].

Stimpson et al. [72] found that most participants (85%) did not seek help from disaster relief services, despite the number of previous floods experienced by the participants. They also found that participants with more previous flood experience had higher odds of seeking flood-related help afterward than those with less previous flood experience ($aOR = 1.58, p < 0.0001$). Participants with higher social support had lower odds of seeking flood-related help afterward than those with a lower amount of social support ($aOR = 0.55, p = 0.011$). Although this study had a large sample size, it must be noted that most participants had minimal flood experience and exposure, and those who did may have fallen out during the follow-up interview. Also, it was unclear if other types of services, such as flood insurance, mitigated the outcome. In spite, this study did find that participants seeking help after a flood due to the social disruption caused by the flood were likely from participants who had more previous flood experience and less social support.

4.3.3.6 Norris et al. [48]

Norris et al. [48] interviewed 658 adults living in two communities in Mexico during the 1999 floods. Based on an earlier publication, interviews started six months after the floods in 2000 and were repeated every six months until 24 months had passed [49]. This study assessed social embeddedness [social disruption], which included a 10-item measure based on subscales of household embeddedness, family embeddedness, and friend embeddedness. Pre-disaster major depressive disorder [health] was tested for its association with social embeddedness. Norris et al. [48] did not find any significant associations between pre-disaster major depressive disorder and social embeddedness or its subscales (friend embeddedness, family embeddedness, and household embeddedness), however, they found gender to have an interactive effect on social embeddedness after the flood. This may be attributed to this study having a more significant proportion of female respondents. Also, it was unclear whether language was a barrier or bias, as consistency of which language was used during the survey and the quality of understanding of the questions are essential to consider. This may have been a factor in why no association was found between the element of health and social disruption.

4.3.4 Outcome of Economic Harm

Only three studies tested the association between an element of community resilience and economic harm. Two of them were carried out in the USA, and the other was conducted in the UK. The community resilience elements seen in these studies were *communication, health, governance and leadership, preparedness, and mental outlook.* Not many studies found in this review have examined the connection between economic harm and an element of community resilience. However, participants who felt weather-specific distress seemed to have a higher likelihood of feeling the losses caused by the floods. No connection was seen between participants who had flood-damage costs and received and acted on a flood warning. Likewise, no connection was seen between participants who had property damage and pre-mental health issues or had spent a long time in a transition camp. Table 4.5 captures the evidence of the associations with economic harm found in each study.

4.3.4.1 Lamond et al. [36]

The first study was carried out in the UK with people affected by the 2007 floods. Lamond et al. [36] tested the association between the actual reinstatement costs

Table 4.5 Evidence table for the outcome of economic harm

Reference(s)	Measures of elements	Identified element	Reported association(s) with economic harm
Norris and Murrell [50], Phifer and Norris 59]	Weather-specific distress	Mental outlook	• Significant association with community destruction caused by the flood ($R^2 = 0.037, p < 0.01$) • A significant interaction was observed with weather-specific distress from participants who had high personal loss and great amount of prior flood exposure ($R^2 = 0.026$, $p < 0.05$) • A significant interaction was observed with weather-specific distress from participants who had high personal loss and great amount of prior trauma experience ($R^2 = 0.044$, $p < 0.01$) Among participants with no prior flood exposure: • Participants with weather-specific distress were found to be associated with higher amount of personal loss and community destruction (respectively, $R^2 = 0.051, p < 0.01$; $R^2 = 0.027, p < 0.05$)
Norris et al. [54]	Pre-Katrina mental health	Health	• No significant association was found with the severity of property damage
	Time in transition camp	Governance and leadership	• No significant association was found with the severity of property damage • A significant interaction was observed between property damage and time in transition camp significant ($\beta = 0.23, p < 0.05$)
Lamond et al. [36]	Receiving a flood warning	Communication	• No significant association was with participant's actual reinstatement costs from the flood impact
	Compliance to warning prior to a flood	Preparedness	• No significant association was with participant's actual reinstatement costs from the flood impact

arising from the flood impact [economic harm] and whether receiving a flood warning [communication] and its compliance prior to a flood [preparedness] made a difference. Actual reinstatement costs were found through participant's insurance claim records of the damage caused by the floods. The questions that matched to *communication* and *preparedness*, respectively, were whether the participant received a flood warning and whether they did anything to prevent damage based on the flood warning. No significant associations were found between reinstatement costs and either element. Still, it is essential to remember that this study included a small and non-representative sample of the UK, and six years had passed between the flood and data collection. Still, one of the strengths of this study was having access to the insurance claim data and using the actual costs of damages on the participants, which may be helpful for a future or follow-up study.

4.3.4.2 Norris and Murrell [50], Phifer and Norris [59]

The second study was carried out in the USA and used a prospective design with five follow-up intervals and two flood occurrences within the study time period. Norris and colleagues [50, 59] examined the personal loss [economic harm] and community destruction [economic harm] caused by the floods in 1981 and 1984. Personal loss was based on participant's perceptions of their losses and impacts on their property and belongings. Community destruction was measured as "the 'impact ratio,' the number of homes in the respondent's county that were damaged or destroyed relative to the population size of that county" ([50], p. 671). The authors tested the association of these outcomes with weather-specific distress [mental outlook], which was measured using four statements to "tap feelings of apprehension during mildly threatening weather," and they were the following: "When it starts raining, I get afraid; I have difficulty sleeping when it rains at night; the weather more or less controls my thoughts; and even during a light rain I wonder whether the river or creek will flood" ([50], p. 670). These questions aligned closer to indicating about participant's attitude and coping skills to a flood than a mental health indicator.

Based on 203 adults, Norris and Murrell [50] found community destruction to be significantly associated with weather-specific distress ($R^2 = 0.037, p < 0.01$). Among participants with no prior flood exposure, the amount of personal loss and community destruction were found to be associated with having weather-specific distress (respectively, $R^2 = 0.051$, $p < 0.01$; $R^2 = 0.027$, $p < 0.05$). Among participants with no prior trauma experience, those with weather-specific distress were found to be associated with higher amount of personal loss and community destruction (respectively, $R^2 = 0.122$, $p < 0.001$; $R^2 = 0.025$, $p < 0.05$). A significant interaction between higher personal loss and a greater amount of prior flood exposure was associated with weather-specific distress ($R^2 = 0.026$, $p < 0.05$). A significant interaction was observed with weather-specific distress among participants who had a high personal loss and a great amount of prior trauma experience ($R^2 = 0.044$, $p < 0.01$). It must be noted that the study may not have sufficient power in sample size, and their sample for participants with no prior flood exposure was 21 people.

As mentioned earlier in psychological impact, this study was poorly explained and unclear with its descriptions of the methodology, measured variables, data collection, and analysis. Still, significant associations were found between the element of mental outlook and economic harm.

4.3.4.3 Norris et al. [54]

In a post-Katrina study by Norris et al. [54], a regression model based on 80 Vietnamese American adults examined the interactions of pre-Katrina mental health [health] and the length of time in transition camps [governance and leaderships] with flood exposure and property damage [economic harm]. Non-significant associations were found in a model consisting of pre-Katrina mental health ($\beta = -0.22$), time in transition camp ($\beta = 0.02$), the severity of property damage and other objective exposure to Hurricane Katrina, and PTSD and other trauma exposure to Hurricane Katrina. However, the interaction between property damage and time in the transition camp was significant ($\beta = 0.23, p < 0.05$). Still, it must be noted that this study may have bias stemming from the unknown validity of the self-developed questions by the authors and may even have recall bias from the participant's self-reported responses. Nonetheless, further examination may be needed about the interaction between property damage and time in the transition camp.

4.3.5 Summary of Associations

In this review, 16 studies reported the relationships between the impacts of flooding with a range of factors that correspond to the nine core elements of what could make a community resilient to flooding. Table 4.6 provides an overview of each association tested, counting only one publication per association, despite some publications reporting more than one. In terms of outcomes, most publications (84%) assessed psychological impact, with fewer evaluating social disruption (31%) and economic harm (21%). The most common element of community resilience assessed was *health* (assessed by 68% of publications), followed by *community networks and relationships* (53%) and *mental outlook* (47%). The included studies did not assess the associations between *local knowledge* or *resources* and NRR security outcomes.

Only the *health* and *mental outlook* elements were studied in relation to each NRR security outcome. Mainly comprising of physical and mental health indicators, the element of *health* was directly associated with each NRR security outcome; if poor health existed for a person before flooding, then a more significant likelihood existed that this person would experience an NRR security outcome. Similarly, having a poor mental outlook was found to be associated with each NRR outcome as well. Surprisingly, only one study [36] assessed the element of *preparedness* beyond more than one NRR security outcome. Lamond et al. [36] only found a significant relationship between *preparedness* and *psychological impact*. Understanding the relationship

Table 4.6 Flooding studies that tested elements of community resilience for association with NRR security outcomes

Elements	National Risk Register Security Outcomes		
	Psychological Impact	Economic Harm	Social Disruption
Local Knowledge			
Community Networks and Relationships	[5]*, [7, 8]*, [13, 25, 26, 55, 74]		[5]*, [12]*, [25]*, [72]*
Communication	[26, 36, 55, 74]*	[36]	
Health	[4]*, [13, 14]*, [26, 55]*, [43]*, [50, 59]*, [54], [74]*	[54]	[14]*, [25, 48],
Governance and Leadership	[5]*, [36]*, [43]*, [54], [74]*	[54]	
Resources			
Economic Investment	[4]*, [26, 55], [36]*, [74]*		
Preparedness	[26, 55]*, [36]*	[36]	
Mental Outlook	[26, 55]*, [29]*, [43, 50, 59]*	[50, 59]*	[72]*

*Study found a significant association

between costs created by flood damage and savings created by preparedness activities would be helpful to empower communities to be ready for a flood. The *local knowledge* and *resources* elements were not tested for their association with NRR security outcomes in the papers identified.

4.4 Discussion

Much research has been carried out on people impacted by flooding, whether the psychological impacts [2, 20, 53, 71, 73, 83], the economic impacts [1, 34, 41, 47], or social impacts [46, 68, 22, 63, 75]. Most of these studies used flooding as an exposure variable and explored this variable's association with their specified outcome.

These studies that report the associations with flood exposure are helpful and provide evidence about the immediate impacts of flooding on psychological, social, and economic outcomes, also identified as the NRR security outcomes. However, they are less useful in identifying possible factors that can mediate or moderate this relationship. Summarized in detail below is how the elements of community resilience relate to the NRR security outcomes.

4.4.1 Elements of Community Resilience and Their Association with the Psychological Impact

Measures identified as psychological impact primarily consisted of checklists and scales concerning traumatic experiences. All the elements of community resilience except *local knowledge* and *resources* were reported to have an association with psychological impact in varying degrees. A trend from the reported associations was found: if people had better *community networks and relationships,* were more engaged in preventative action [*preparedness*], and had no pre-existing *health* issues, they were less likely to have a severe psychological impact from a flood. Keeping a positive *mental outlook* was also a mediating variable to reduce psychological impact. The measures related to *communication, economic investment,* and *governance and leadership* were associated with psychological impact, but the identified factors only included some aspects of these elements. For example, higher psychological impact was likely found among participants who rented rather than owned their property [4], but *economic investment* was defined much larger than the renting status of individuals. The financial measures found after the flood were seen as an outcome of economic harm, which included financial loss, strain, or difficulties caused by the flood. The measures of financial loss, strain, or difficulties were reported to be associated with psychological impact as well [5, 13, 14].

4.4.2 Elements of Community Resilience and Their Association with Social Disruption

Within the six studies that tested an association between social disruption and an element of community resilience, the identified measures varied from adverse event experiences from the flood to thwarted belongingness after the flood. The common denominator for the outcome was that flooding disrupted the participants' daily lives in a significant way. With more than one study providing results, the elements of community resilience associated with social disruption were *community networks and relationships, health, and economic investment.* Despite only identifying six studies, measures related to these elements reduced social disruption among flood-affected participants. For example, people with higher social support [5, 72], better access to medical care [14], and fewer economic difficulties [economic harm] [14, 72] were found to be less likely to have a social disruption.

4.4.3 *Elements of Community Resilience and Their Association with Economic Harm*

Economic harm measured in the reviewed studies focuses on the damages caused to participants and property by flooding. Only three studies were found, and they were carried out in the USA or UK. It is unclear whether a significant association exists between economic harm and *communication, preparedness, and mental outlook.* Only one study was identified for each element. Only one significant association was found among these elements, which was a factor related to *mental outlook.*

Learning more about reducing economic harm in flood-affected communities is crucial if a better understanding of how the elements of community resilience can be applied in communities is to be reached. Still, practical implications for economic harm with respects to flooding were found in previous research, which focused on the cost of flood damage [38], the built environment in the communities and the impact on property value [39], and the costs of mitigation measures, typically presented as case studies [35]. For example, Lamond and Proverbs [37] saw a short-term effect (less than three years) of house prices falling after the flood due to not keeping up with the growth in the market, as seen in a case study of the village of Barlby in North Yorkshire, UK. Although this is one case study, further quantitative studies are needed to identify additional links between the elements of community resilience measured before a flood that can mitigate economic harm after a flood.

4.4.4 *Absence of Elements of Community Resilience and Its Association with NRR Security Outcomes and Flooding*

The absence of studies looking at *local knowledge and resources* factors does not mean that there is no connection with NRR security outcomes. This review has demonstrated how limited the flooding literature is regarding these elements and their potential associations with a NRR security outcome. Previous research studies have shown the importance of *local knowledge* when it comes to flooding. For example, in an analysis of coastal flooding in Quebec, Canada by Bernatchez et al. [6], the authors discussed the importance of *local knowledge* in increasing the understanding of flood patterns and potential flood disruption and levels that could occur. Additionally, prior research has highlighted the importance of *local knowledge* regarding disasters and their management more generally [3, 15, 44, 62]. For example, Mercer et al. [44] stressed the importance of creating an integrated framework between indigenous and scientific knowledge when it comes to reducing the community's vulnerability to disasters. Comfort and Okada [15] suggested the development of "Knowledge Commons," which is described as a forum where information can be quickly and continually searched, exchanged, and updated between individuals, groups and/or organizations. The goal of having this type of forum is to improve decision-making when dealing with extreme events under distress.

The element of *resources* has been researched and found to be important for flooding. Specific examples of this from past studies, which did not meet this review's selection criteria, were on community's water supplies after severe flooding in Sri Lanka [21], household and individual sanitation after a flood, such as hygienic bathroom facility and practices [46], and establishing better food security during floods in south Asia [19]. One of the studies included in this review also discussed the importance of *mental outlook*, trauma *health*, and *resources*. Stimpson et al. ([72], p. 140) stated that experiencing trauma "lowers the capacity for individuals to cope with stress by reducing the psychosocial resources available to combat stress." The authors applied this understanding to disaster relief services. They concluded that "additional floods will tap into an increasing lower supply of resources until individuals may have no other resource other than to seek outside assistance" ([72], pp. 140–141).

The absence of studies examining the missing elements of community resilience with NRR security outcomes does not mean nor indicate that those elements are not crucial in a community facing a flood. Both *local knowledge* and *resources* have been known to improve the community and enhance other aspects of community resilience in the process. Until further studies can measure these impacts and their role in mitigating damage caused by flooding, the list of elements, derived from the meanings of community resilience, should not be discounted as not relevant to flood-affected communities.

4.4.5 Limitations of the Reviewed Studies

Several limitations need consideration for the studies included in this review. First, most of the studies were cross-sectional in design, so causality cannot be inferred. Accurate and validated measures for elements of community resilience were challenging to find. Most authors either created their own set of questions or a single question, such as "do you have a regular doctor or a clinic where you can go for your routine medical care" to describe participant's access to medical care ([14], p. 418). Many studies had a lengthy time difference between the occurrence of a flood and data collection, which could lead to recall bias. Over 60% of the studies analyzed a sample size less than 600 people. Of which, most were cross-sectional studies. Such studies below that sample size may not have sufficient statistical power to explore associations. The generalizability of the findings from the studies was another limitation. Most were specific to a location and a flood or did not have a demographically representative sample reflecting the study participants' country.

4.4.6 Limitations and Implications of This Chapter

Findings in this chapter need to be interpreted alongside the challenges faced in this review and its design. For example, two major issues occurred when categorizing past studies' measures as either an element of community resilience or a NRR security outcome. One was that measures could be argued to various elements or outcomes, depending on the context. For example, Gordon et al. [25] examined the relationship between perceived thwarted belongingness and perceived burdensomeness. Both measures described the outcome of perceived isolation [social disruption] that participants felt due to flooding. However, when tested against each other, perceived burdensomeness, the level of feeling cumbersome upon others, could be seen as *community networks and* relationships. This variable was ultimately decided to measure the feeling of being cumbersome upon others *after the flood*. Without the flood, this variable would not be measured; thus, both were labeled as outcomes of social disruption. This type of confusion was limited in this review, hence why the marking criteria were definitive that only one variable was seen as an element of community resilience or NRR outcome in one association in each study. That label does not change in the study, even if it is then tested in a different association within the same study.

However, this leads to another limitation of this review. In order to make the methodology of this review systematic and clear, the elements of community resilience were labeled as measures that were not direct outcomes of flooding, which were left for NRR security outcomes. As seen in Patel et al. [57] and Chap. 1, publications, such as ones by Bonanno [9] and Gibson [23], viewed and applied community resilience as an outcome of experiencing a disaster. The overall approach used in this chapter prioritized identifying past studies' outcome measures to match with the NRR security outcomes instead of the community resilience elements.

As a result, Table 4.7 shows the studies in this review with associations reported between variables identified as NRR security outcomes. Some of these variables could be matched to an element of community resilience. Therefore, Table 4.7 is essential to consider in conjunction with the findings found in this chapter. More studies were seen to identify associations between NRR security outcomes than what was shared in the results of this review. This could have been due to the nature and

Table 4.7 Studies reporting associations between NRR security outcomes

National Risk Register Security Outcomes	National Risk Register Security Outcomes		
	Psychological Impact	Economic Harm	Social Disruption
Psychological Impact	[7, 8]*, [54]*	[5]*, [13]*, [14]*, [26, 55]*, [54]*, [50, 59]*	[14]*, [25, 26, 55]*
Economic Harm			[14]*, [72]*
Social Disruption			[25]*

*Study found a significant association

methodology of the studies found in this review which reported more on outcomes found than possible attributes.

Nevertheless, six studies reported associations between economic harm and psychological impact. Contrastingly, this was more studies than the entire amount found in this review reporting an association with economic harm and any element of community resilience. It is important to note this discrepancy because more associations between economic harm and an element could exist if the marking methodology differed. Future studies looking to develop the understanding of community resilience further should explore these types of measures and how the elements can be better identified and measured against NRR security outcomes.

The nature of community resilience, where labels could change based on the perspective of the reviewer or author, was a challenge. This review's methodology strictly adhered to a systematic and logical approach to limit the challenge. Labels used for this review could be quite different if community resilience was seen and measured as a process, where elements of community resilience exist before, during, and after flooding. The outcomes-to-outcomes associations in Table 4.7 were intended to illustrate this. Plausible variables or relationships could be mapped as elements of community resilience or NRR security outcomes if interpreted differently from included studies. However, in this review, a systematic and logical approach to the labeling was best described in the methodology to limit and explain any ambiguity that may have been seen when marking each measure as an element or a NRR security outcome.

Another limitation was the strength of comparability for associations found. Due to the low number of studies, conclusions based on these associations need further research to confirm if an association was present between an element of community resilience and a NRR security outcome. Associations do not show directionality, and this review used the original author's selection in the test to interpret direction. Additionally, it may be possible that some aspects of a community resilience element are linked to NRR security outcomes while others are not.

In the same way that community resilience is too broad of a term to be useful, the words making up the nine elements of community resilience, such as *economic investment* or *governance and leadership*, may also be too broad to be identifiable and measured fairly. This broadness created an issue in identifying and labeling measures in the review. There may have been too inclusive and generous with accepting variables, which may lead to over-reporting for an element. Future studies will need to be done to add details and narrow down the aspects of community resilience to understand the driving force of why and how they mitigate the harm caused by flooding.

Selection bias could have occurred. This review was carried out with a sole researcher identifying the studies and marking the measures found. Limiting this bias was done by systematically searching for the studies and labeling the measures per the criteria given in the methods. The implication of this could be that studies may be missing or extra studies may be added that may not necessarily fit with the understanding of elements of community resilience and NRR security outcomes.

This implication also applies to the relatively few studies found for each element of community resilience. Lacking many studies for each association, further work is needed to confirm whether elements of community resilience influence the NRR security outcomes seen in communities. This review did not find the same type of questionnaire or measures used across the studies either. Also, the original author's intentions or descriptions in the studies were not always clear. This could have been improved by contacting the original authors directly to receive clarification on their research which would increase the accuracy of the studies being marked as an element of community resilience or NRR security outcome.

As seen in this review, existing associations show a reduction in harm caused by flooding due to a positive influence of an element of community resilience. However, not all measures identified as an element were tested nor found to be significantly associated with a NRR security outcome. Future studies can help clarify how community resilience improves a community's chances for less observed impact caused by severe flooding.

4.5 Conclusion

In this Chapter, the aim was to examine the associations between the elements of community resilience and NRR outcomes in previous flooding studies. From 19 publications, 16 studies were seen to test the association between the elements of community resilience and NRR outcomes in flood-affected populations. The majority of the included publications (84%) assessed the association of elements of community resilience with psychological impact, while fewer assessed their association with social disruption (31%) or economic harm (21%). These studies show some evidence that elements of community resilience can reduce the harm caused by flooding, but further research needs to be carried out to have a deeper understanding and clarity of these associations. Future studies can clarify whether to treat the elements of community resilience as predictors of outcomes and when best to apply the concept in the disaster cycle. The findings of this Chapter add to a broader understanding of community resilience and its ability to enhance a community's response to flooding, by identifying existing studies that tested the relationships between elements of community resilience and harmful security and health outcomes.

References

1. Aerts JC, Botzen WJ, de Moel H, Bowman M (2013) Cost estimates for flood resilience and protection strategies in New York City. Ann N Y Acad Sci 1294:1–104. https://doi.org/10.1111/nyas.12200
2. Ahern M, Kovats RS, Wilkinson P, Few R, Matthies F (2005) Global health impacts of floods: epidemiologic evidence. Epidemiol Rev 27(1):36–46. https://doi.org/10.1093/epirev/mxi004

3. Alcántara-Ayala I (2004) Flowing mountains in mexico. Mt Res Dev 24(1):10–13. https://doi.org/10.1659/0276-4741(2004)024[0010:FMIM]2.0.CO;2
4. Alderman K, Turner LR, Tong S (2013) Assessment of the health impacts of the 2011 summer floods in Brisbane. Disaster Med Public Health Prep 7(4):380–386. https://doi.org/10.1017/dmp.2013.42
5. Bei B, Bryant C, Gilson KM, Koh J, Gibson P, Komiti A et al (2013) A prospective study of the impact of floods on the mental and physical health of older adults. Aging Ment Health 17(8):992–1002. https://doi.org/10.1080/13607863.2013.799119
6. Bernatchez P, Fraser C, Lefaivre D, Dugas S (2011) Integrating anthropogenic factors, geomorphological indicators and local knowledge in the analysis of coastal flooding and erosion hazards. Ocean Coast Manag 54(8):621–632. https://doi.org/10.1016/j.ocecoaman.2011.06.001
7. Bokszczanin A (2007) PTSD symptoms in children and adolescents 28 months after a flood: age and gender differences. J Trauma Stress 20(3):347–351. https://doi.org/10.1002/jts.20220
8. Bokszczanin A (2008) Parental support, family conflict, and overprotectiveness: predicting PTSD symptom levels of adolescents 28 months after a natural disaster. Anxiety Stress Coping 21(4):325–335. https://doi.org/10.1080/10615800801950584
9. Bonanno GA (2004) Loss, trauma, and human resilience: have we underestimated the human capacity to thrive after extremely aversive events? Am Psychol 59(1):20–28. https://doi.org/10.1037/0003-066x.59.1.20
10. Buckner JC (1988) The development of an instrument to measure neighborhood cohesion. Am J Community Psychol 16(6):771–791. https://doi.org/10.1007/BF00930892
11. Cabinet Office (2015) National risk register for civil emergencies-2015 edition. London, UK. https://www.gov.uk/government/uploads/system/uploads/attachment_data/file/419549/20150331_2015-NRR-WA_Final.pdf
12. Chang K (2010) Community cohesion after a natural disaster: insights from a Carlisle flood. Disasters 34(2):289–302. https://doi.org/10.1111/j.03613666.2009.01129.x
13. Chen AC, Keith VM, Leong KJ, Airriess C, Li W, Chung KY et al (2007) Hurricane Katrina: prior trauma, poverty and health among Vietnamese-American survivors. Int Nurs Rev 54(4):324–331. https://doi.org/10.1111/j.1466-7657.2007.00597
14. Collins TW, Jimenez AM, Grineski SE (2013) Hispanic health disparities after a flood disaster: results of a population-based survey of individuals experiencing home site damage in El Paso (Texas, USA). J Immigr Minor Health 15(2):415–426. https://doi.org/10.1007/s10903-012-9626-2
15. Comfort LK, Okada A (2013) Emergent leadership in extreme events: a knowledge commons for sustainable communities. Int Rev Public Administr 18(1):61–77. https://doi.org/10.1080/12294659.2013.10805240
16. Davis TE, Grills-Taquechel AE, Ollendick TH (2010) The psychological impact from Hurricane Katrina: effects of displacement and trauma exposure on university students. Behav Ther 41(3):340–349. https://doi.org/10.1016/j.beth.2009.09.004
17. DeBono R, Vincenti K, Calleja N (2012) Risk communication: climate change as a human-health threat, a survey of public perceptions in Malta. Eur J Pub Health 22(1):144–149. https://doi.org/10.1093/eurpub/ckq181
18. Diener E, Emmons RA, Larsen RJ, Griffin S (1985) The satisfaction with life scale. J Pers Assess 49(1):71–75. https://doi.org/10.1207/s15327752jpa4901_13
19. Douglas I (2009) Climate change, flooding and food security in south Asia. Food Security 1(2):127–136. https://doi.org/10.1007/s12571-009-0015-1
20. Du W, FitzGerald GJ, Clark M, Hou XY (2010) Health impacts of floods. Prehosp Disaster Med 25(3):265–272. https://doi.org/10.1017/S1049023X00008141
21. Ferretti E, Bonadonna L, Lucentini L, Della Libera S, Semproni M, Ottaviani M (2010) A case study of sanitary survey on community drinking water supplies after a severe (post-Tsunami) flooding event. Ann Ist Super Sanita 46:236–241. https://doi.org/10.1590/S0021-25712010000300003

22. Fundter DQ, Jonkman B, Beerman S, Goemans CL, Briggs R, Coumans F et al (2008) Health impacts of large-scale floods: governmental decision-making and resilience of the citizens. Prehospital Disaster Med 23(4):s70–73. https://doi.org/10.1017/S1049023X00021282
23. Gibson CA (2010) An integrated approach to managing disruption-related risk: Life and death in a model community. J Bus Contin Emerg Plan 4(3):246–261. Retrieved from http://www.ingentaconnect.com/content/hsp/jbcep/2010/00000004/00000003/art00006
24. Goldberg D, Williams P (1988) A user's guide to the General Health Questionnaire. Windsor, UK: NFER Nelson Retrieved from https://www.worldcat.org/title/users-guide-to-the-general-health-questionnaire/oclc/26545847.
25. Gordon KH, Bresin K, Dombeck J, Routledge C, Wonderlich JA (2011) The impact of the 2009 red river flood on interpersonal risk factors for suicide. Crisis: J Crisis Interv Suicide 32(1):52–55. https://doi.org/10.1027/0227-5910/a000051
26. Greene G, Paranjothy S, Palmer SR (2015) Resilience and vulnerability to the psychological harm from flooding: the role of social cohesion. Am J Public Health 105(9):1792–1795. https://doi.org/10.2105/ajph.2015.302709
27. Hartz DA, Golden JS, Sister C, Chuang WC, Brazel AJ (2012) Climate and heat-related emergencies in Chicago, Illinois (2003–2006). Int J Biometeorol 56(1):71–83. https://doi.org/10.1007/s00484-010-0398-x
28. Hertzog C, Van Alstine J, Usala PD, Hultsch DF, Dixon R (1990) Measurement properties of the Center for Epidemiological Studies Depression Scale (CES-D) in older populations. Psychol Assess J Consul Clinical Psychol 2(1):64–72. https://doi.org/10.1037/1040-3590.2.1.64
29. Huang P, Tan H, Liu A, Feng S, Chen M (2010) Prediction of posttraumatic stress disorder among adults in flood district. BMC Public Health 10:207. https://doi.org/10.1186/1471-2458-10-207
30. Joseph R, Proverbs D, Lamond J (2014) Resilient reinstatement: what can we learn from the 2007 flooding in England? Flood Recovery. Innov Response IV 184:175. https://doi.org/10.2495/FRIAR140151
31. Kessler RC, Barker PR, Colpe LJ et al (2003) Screening for serious mental illness in the general population. Arch Gen Psychiatry 60(2):184–189. https://doi.org/10.1001/archpsyc.60.2.184
32. Kjellstrom T, McMichael AJ (2013) Climate change threats to population health and well-being: the imperative of protective solutions that will last. Glob Health Action 6:20816. https://doi.org/10.3402/gha.v6i0.20816
33. Kroenke K, Spitzer RL, Williams JB (2001) The PHQ-9: validity of a brief depression severity measure. J Gen Intern Med 16(9):606–613
34. Lam NS, Arenas H, Pace K, LeSage J, Campanella R (2012) Predictors of business return in New Orleans after Hurricane Katrina. PLOS ONE [Electronic Resource] 7(10):e47935. https://doi.org/10.1371/journal.pone.0047935
35. Lamond J (2011) Financial implications of flooding and the risk of flooding on households. In: Lamond J, Booth C, Hammond F, Proverbs D (Eds), Flood hazards: impacts and responses for the built environment. CRC Press, pp 317–326. https://doi.org/10.1201/b11050-28
36. Lamond J, Joseph RD, Proverbs DG (2015) An exploration of factors affecting the long term psychological impact and deterioration of mental health in flooded households. Environ Res 140:325–334. https://doi.org/10.1016/j.envres.2015.04.008
37. Lamond J, Proverbs D (2006) Does the price impact of flooding fade away? Struct Surv 24(5):363–377. https://doi.org/10.1108/02630800610711960
38. Lamond J, Proverbs D, Antwi A (2005) The effect of floods and floodplain designation on the value of property: an analysis of past studies. Paper presented at the 2nd Probe Conference, Glasgow, Scotland. Retrieved from https://www.irbnet.de/daten/iconda/CIB10671.pdf
39. Lamond J, Proverbs D, Hammond F (2010) The impact of flooding on the price of residential property: a transactional analysis of the UK market. Hous Stud 25(3):335–356. https://doi.org/10.1080/02673031003711543
40. Lang AJ, Stein MB (2005) An abbreviated PTSD checklist for use as a screening instrument in primary care. Behav Res Ther 43(5):585–594. https://doi.org/10.1016/j.brat.2004.04.005

41. Leiter AM, Oberhofer H, Raschky PA (2009) Creative disasters? Flooding effects on capital, labour and productivity within European firms. Environ Resource Econ 43(3):333–350. https://doi.org/10.1007/s10640-009-9273-9
42. Leppämäki S, Meesters Y, Haukka J, Lönnqvist J, Partonen T (2003) Effect of simulated dawn on quality of sleep—a community-based trial. BMC Psychiatry 3(1):1–5. https://doi.org/10.1186/1471-244x-3-14
43. Mason V, Andrews H, Upton D (2010) The psychological impact of exposure to floods. Psychol Health Med 15(1):61–73. https://doi.org/10.1080/13548500903483478
44. Mercer J, Kelman I, Taranis L, Suchet-Pearson S (2010) Framework for integrating indigenous and scientific knowledge for disaster risk reduction. Disasters 34(1):214–239. https://doi.org/10.1111/j.1467-7717.2009.01126.x
45. Mollica RF, Caspi-Yavin Y, Bollini P, Truong T, Tor S, Lavelle J (1992) The Harvard Trauma Questionnaire. Validating a cross-cultural instrument for measuring torture, trauma, and post-traumatic stress disorder in Indochinese refugees. J Nervous Mental Dis 180(2):111–116. Retrieved from https://www.ncbi.nlm.nih.gov/pubmed/1737972
46. Morgan O, Ahern M, Cairncross S (2005) Revisiting the tsunami: health consequences of flooding. PLoS Med 2(6):e184. https://doi.org/10.1371/journal.pmed.0020184.g001
47. Navrud S, Tuan TH, Tinh BD (2012) Estimating the welfare loss to households from natural disasters in developing countries: a contingent valuation study of flooding in Vietnam. Glob Health Action 5:17609. https://doi.org/10.3402/gha.v5i0.17609
48. Norris FH, Baker CK, Murphy AD, Kaniasty K (2005) Social support mobilization and deterioration after Mexico's 1999 flood: Effects of context, gender, and time. Am J Community Psychol 36(1–2):15–28. https://doi.org/10.1007/s10464-005-6230-9
49. Norris FH, Murphy AD, Baker CK, Perilla JL (2004) Postdisaster PTSD over four waves of a panel study of Mexico's 1999 flood. J Trauma Stress 17(4):283–292. https://doi.org/10.1023/B:JOTS.0000038476.87634.9b
50. Norris FH, Murrell SA (1988) Prior experience as a moderator of disaster impact on anxiety symptoms in older adults. Am J Community Psychol 16(5):665–683. https://doi.org/10.1007/BF00930020
51. Norris FH, Perilla JL (1996) The revised Civilian Mississippi Scale for PTSD: reliability, validity, and cross-language stability. J Trauma Stress 9(2):285–298. https://doi.org/10.1002/jts.2490090210
52. Norris FH, Stevens SP, Pfefferbaum B, Wyche KF, Pfefferbaum RL (2008) Community resilience as a metaphor, theory, set of capacities, and strategy for disaster readiness. Am J Community Psychol 41(1–2):127–150. https://doi.org/10.1007/s10464-007-9156-6
53. Norris FH, Tracy M, Galea S (2009) Looking for resilience: understanding the longitudinal trajectories of responses to stress. Soc Sci Med 68(12):2190–2198. https://doi.org/10.1016/j.socscimed.2009.03.043
54. Norris FH, Vanlandingham MJ, Vu L (2009) PTSD in Vietnamese Americans following Hurricane Katrina: prevalence, patterns, and predictors. J Trauma Stress 22(2):91–101. https://doi.org/10.1002/jts.20389
55. Paranjothy S, Gallacher J, Amlôt R, Rubin GJ, Page L, Baxter T et al (2011) Psychosocial impact of the summer 2007 floods in England. BMC Public Health 11:145. https://doi.org/10.1186/1471-2458-11-145
56. Parloff MB, Kelman HC, Frank JD (1954) Comfort, effectiveness, and self-awareness as criteria of improvement in psychotherapy. Am J Psychiatry 111(5):343–352. https://doi.org/10.1176/ajp.111.5.343
57. Patel SS, Rogers MB, Amlôt R, Rubin GJ (2017) What do we mean by 'community resilience'? A systematic literature review of how it is defined in the literature. PLoS Curr 9
58. Perkins DD, Florin P, Rich RC, Wandersman A, Chavis DM (1990) Participation and the social and physical environment of residential blocks: crime and community context. Am J Community Psychol 18(1):83–115. https://doi.org/10.1007/BF00922690
59. Phifer JF, Norris FH (1989) Psychological symptoms in older adults following natural disaster: nature, timing, duration, and course. J Gerontol 44(6):S207–217. https://doi.org/10.1093/geronj/44.6.S207

60. Qureshi S (1996) Minding morale of institutional markets. J Hosp Mark 10(2):37–47. https://doi.org/10.1300/J043v10n02_05
61. Rapley M, Beyer S (1996) Daily activity, community participation and quality of life in an ordinary housing network. J Appl Res Intellect Disabil 9(1):31–39. https://doi.org/10.1111/j.1468-3148.1996.tb00096.x
62. Raymond CM, Fazey I, Reed MS, Stringer LC, Robinson GM, Evely AC (2010) Integrating local and scientific knowledge for environmental management. J Environ Manage 91(8):1766–1777. https://doi.org/10.1016/j.jenvman.2010.03.023
63. Reifels L, Bassilios B, Spittal MJ, King K, Fletcher J, Pirkis J (2015) Patterns and predictors of primary mental health service use following bushfire and flood disasters. Disaster Med Public Health Prep 9(3):275–282. https://doi.org/10.1017/dmp.2015.23
64. Risk and Policy Analysts Ltd., Flood Hazard Research Centre 2004, & Defra (2003) The appraisal of human-related intangible impacts of flooding: R&D Technical Report FD2005/TR. London, UK: Defra Flood Management Division
65. Robinson D, Wilkinson D (1995) Sense of community in a remote mining town: Validating a neighborhood cohesion scale. Am J Community Psychol 23(1):137–148. https://doi.org/10.1007/BF02506926
66. Roger D, Jarvis G, Najarian B (1993) Detachment and coping: the construction and validation of a new scale for measuring coping strategies. Personality Individ Differ 15(6):619–626. https://doi.org/10.1016/0191-8869(93)90003-L
67. Scott W, Dua J (1999) Development of a scale to assess posttraumatic stress disorder. Int J Stress Manag 6(3):149–165. https://doi.org/10.1023/a:1021966224338
68. Smith BW, Freedy JR (2000) Psychosocial resource loss as a mediator of the effects of flood exposure on psychological distress and physical symptoms. J Trauma Stress 13(2):349–357. https://doi.org/10.1023/A:1007745920466
69. Spielberger C, Gorsuch R, Lushene R (1970) STAI manual for the State-Trait Anxiety Inventory. Palo Alto, CA: Consulting Psychologists, Inc. Retrieved from http://www.statisticssolutions.com/state-trait-anxiety-inventory-stai/
70. Spitzer RL, Kroenke K, Williams JB, Lowe B (2006) A brief measure for assessing generalized anxiety disorder: the GAD-7. Arch Intern Med 166(10):1092–1097. https://doi.org/10.1001/archinte.166.10.1092
71. Stanke C, Murray V, Amlôt R, Nurse J, Williams R (2012) The effects of flooding on mental health: outcomes and recommendations from a review of the literature. PLOS Curr Disasters 4. https://doi.org/10.1371/4f9f1fa9c3cae
72. Stimpson JP, Wilson FA, Jeffries SK (2008) Seeking help for disaster services after a flood. Disaster Med Public Health Prep 2(3):139–141. https://doi.org/10.1097/DMP.0b013e318183cfbb
73. Tapsell S, Tunstall S, Priest S (2009) Developing a conceptual model of flood impacts upon human health. (T10–07–10). Wallingford, UK: FLOODsite Consortium. Retrieved from http://www.floodsite.net/html/partner_area/project_docs/T10-09-02_Health_Impacts__Model_v1_1_Pn10.pdf
74. Tunstall S, Tapsell S, Green C, Floyd P, George C (2006) The health effects of flooding: social research results from England and Wales. J Water Health 4(3):365–380. https://doi.org/10.2166/wh.2006.031
75. Turner LR, Alderman K, Huang C, Tong S (2013) Impact of the 2011 Queensland floods on the use of tobacco, alcohol and medication. Aust N Z J Public Health 37(4):396. https://doi.org/10.1111/1753-6405.12093
76. Van Orden KA, Witte TK, Gordon KH, Bender TW, Joiner TE Jr (2008) Suicidal desire and the capability for suicide: tests of the interpersonal-psychological theory of suicidal behavior among adults. J Consult Clin Psychol 76(1):72–83. https://doi.org/10.1037/0022-006x.76.1.72
77. Ware JE Jr, Kosinski M, Keller SD (1996) A 12-Item short-form health survey: construction of scales and preliminary tests of reliability and validity. Med Care 34(3):220–233. https://doi.org/10.1097/00005650-199603000-00003

78. Ware JE Jr, Sherbourne CD (1992) The MOS 36-item short-form health survey (SF-36). I. Conceptual framework and item selection. Med Care 30(6):473–483. Retrieved from https://www.ncbi.nlm.nih.gov/pubmed/1593914
79. Warren JA, Berner JE, Curtis T (2005) Climate change and human health: infrastructure impacts to small remote communities in the north. Int J Circumpolar Health 64(5):487–497. https://doi.org/10.1061/40792(173)303
80. Weathers F, Litz B, Herman D, Huska J, Keane T (1994) The PTSD checklist-civilian version (PCL-C). Boston, MA: National Center for PTSD. Retrieved from http://www.ptsd.va.gov/professional/assessment/adult-sr/ptsd-checklist.asp
81. Weems CF, Watts SE, Marsee MA, Taylor LK, Costa NM, Cannon MF et al (2007) The psychosocial impact of Hurricane Katrina: contextual differences in psychological symptoms, social support, and discrimination. Behav Res Ther 45(10):2295–2306. https://doi.org/10.1016/j.brat.2007.04.013
82. Weiss SD, Marmar RC (1997) The impact of event scale-revised. In: Wilson JP, Keane TM (eds) Assessing psychological trauma and PTSD: a handbook for practitioners. Gilford Press, New York, pp 399–411
83. Wind TR, Joshi PC, Kleber RJ, Komproe IH (2013) The impact of recurrent disasters on mental health: a study on seasonal floods in northern India. Prehosp Disaster Med 28(3):279–285. https://doi.org/10.1017/S1049023X13000290
84. World Health Organization (1997) Composite International diagnostic interview (CIDI). Version 2.1. Geneva: World Health Organization Retrieved from http://medicine.yale.edu/intmed/vacs/instruments/CIDI_Lifetime_Paper_Version_WHO-1_tcm75-3299_tcm75-284-32.pdf

Chapter 5
Lessons Learned and Cautious Future Directions When Using Community Resilience Approaches

Abstract This concluding Chapter discusses lessons learned in this book to reduce security and health impacts caused by severe flooding through community resilience approaches. Different groups—researchers, policy-makers, government and community leaders, flood-affected and flood-risk participants saw community resilience differently. From reviewing existing definitions of community resilience, no overarching definition was found among the researchers. Still, nine common themes were identified for how communities can be resilient against disasters. The nine community resilience elements found were: local knowledge, community networks and relationships, communication, health, governance and leadership, resources, economic investment, preparedness, and mental outlook. Using the context of flooding in the United Kingdom, the themes associated with community resilience found by researchers were further elaborated by flood-affected participants. While researchers described macro-level attributes and processes in the community and its emergency management, the flood-affected participants naturally described activities and actions that their community had or needed to be better for the next severe flood. However, a less fine-grained understanding of community resilience was observed among people living in high flood-risk areas. Future research and community programming directions are explored, and the summary insights developed throughout the Book are shared as well as implications for other disasters that require humanitarian and community aid.

Keywords Community resilience · Disasters · Flooding · Disaster preparedness · Emergency management · Public health · Global health · Lessons learned · Future directions · Operationalization · Disaster risk reduction

5.1 Overview

The way disaster-response professionals, government officials, and academics define community resilience affects how to understand and help communities and measure successful outcomes. With more funding and emphasis on disaster risk reduction projects that focus on community resilience, the confusion surrounding the definition

S. S. Patel, *Community Resilience When Disaster Strikes*, Advanced Sciences and Technologies for Security Applications,
https://doi.org/10.1007/978-3-031-07992-4_5

and operationalization of community resilience is troubling. Community resilience is in danger of being used more as a 'buzzword' than an actionable concept with harm-reducing programs built for communities facing disasters.

This book aims to present and discuss research to understand better how to operationalize community resilience to reduce security and health harm caused by severe environmental hazards. In Chaps. 2 and 3, perceptions of both those affected by a severe environmental hazard and those who have not but are at-risk reveal the importance of trust, community resources, and neighborhood security. These learnings were found through the context of flooding in the United Kingdom (UK), which recent projections suggest a higher frequency of river flooding and severe impacts across the UK [20]. Combining evidence from literature reviews, discussions from focus groups and interviews, and results from a cross-sectional survey, this Book offers practical ways of bringing communities together after an environmental hazard to prepare for the next one.

Table 5.1 shows an overview of the analysis of these findings and areas that require further consideration and research. Despite the ambiguities, common elements identified by Patel et al. [32] were examined in practice to see if they could reduce the likely harm caused by a disaster. Findings from qualitative and quantitative research studies shared in this Book explored how community resilience, as a concept, could be interpreted and communicated. The implications were that disaster-response professionals, government officials, and academics operating in flood risk communities would use these findings to enhance their ongoing work.

From a knowledge translation perspective, the results have shown that community resilience as it applies to disasters is understood differently whether it be among the

Table 5.1 Overview of the findings from this book and further aspects that need consideration

	What we do understand	What needs consideration
Definition of community resilience [2]	Range of differing interpretations but three general definitions and their blend: • 'Process' (i.e., an ongoing process of change and adaptation) • 'Absence of adverse effect' (i.e., an ability to maintain stable functioning) • 'Range of attributes' (i.e., a broad collection of response-related abilities) • Blend of more than one above	• Achieving consensus on a definition of 'community resilience' may not help communities at risk of future disasters • Attention to the constituent elements of community resilience will better help communities prepare for disasters
Elements of community resilience in the disaster literature [2]	Nine broad elements emerged from the existing definitions of community resilience: • Local knowledge, community networks and relationships, communication, health, governance and leadership, resources, economic investment, preparedness, and mental outlook	• Strategies to maximize the number of activities addressing each element for communities • Identify at what time point (i.e., pre-, during, and post-disaster setting) should these elements be most effectively applied

(continued)

Table 5.1 (continued)

	What we do understand	What needs consideration
Elements of community resilience as told by flood-affected UK residents (Chap. 2)	• Residents stressed the importance of activities related to the elements and gave specifics about which were useful or not • Some details, such as knowing their neighbors or who to contact locally for help, were said to be most useful if established in the pre-disaster setting • Other specifics, such as dealing with insurance and helping with uncertainty in the community, were said to be useful if addressed in the post-disaster setting • Overall, residents confirmed that the elements helped or would have helped reduce the amount of harm they faced in the community • Many residents asked for more attention from officials to root successful examples of the elements in their communities	• Understanding how the elements can best be applied and rooted into the community before a disaster needs further clarity and support • A community's culture may be different within or across localities. Future studies should examine how culture influences the elements of community resilience and their impact on NRR security outcomes
Associations between elements of community resilience and NRR outcomes, as determined by UK residents living in high-risk areas for flooding (Chap. 3)	• Residents understood elements of community resilience as positive connotations to three general categories: strong fellowship in the community, trust in the people in charge or system in place, and memories of past disasters • Residents understood NRR security outcomes as negative connotations to two general categories: impacts at a personal level and impacts at a community level • Participants with a stronger belief in fellowship in the community perceived a flood would cause more likelihood of personal impact and less community impact • Participants who had higher trust in the system perceived a flood would have less likelihood of community impact • Participants who felt a greater sense of belonging to the community or knew or guessed the local disaster support center's location perceived a flood would cause less community impact • Participants with more excellent knowledge of disasters perceived a flood would cause less personal impact and more community impact	• This new understanding of the elements of community resilience and their impact on the NRR security outcomes in high flood-risk areas in the UK should be integrated into current efforts in disaster preparedness and risk communication in these communities • Developing an accurate and consistent measurement scale for each element and each NRR security outcome will help to give a complete picture of the associations between the elements and NRR security outcomes in research carried out pre- and post-flooding

(continued)

Table 5.1 (continued)

	What we do understand	What needs consideration
Elements of community resilience in the flood-specific literature (Chap. 4)	• Few previous studies look at associations between the elements and psychological impact, economic harm, or social disruption • The quality of existing studies is mixed	• Future quantitative studies should explore the associations between elements of community resilience and NRR security outcomes in flood-affected communities • Studies should seek to confirm the linkages between the elements of community resilience and other factors that can enhance a community against a flood

scientific community, by flood victims in Kent and Surrey, and by people living in high flood-risk areas in the UK. The gap in the scientific literature was based on ambiguity, limiting the application of the concept to be useful to aid communities. However, nine common elements were found in these definitions: local knowledge, community networks and relationships, communication, health, governance and leadership, resources, economic investment, preparedness, and mental outlook. These were all identified as being important during all stages of a severe flood, with some being more prevalent in preparation, during, or afterward dealing with the aftermath of an environmental disaster.

Previous flooding studies demonstrated a relationship between these elements of community resilience and the outcomes of psychological impact, economic harm, and social disruption (NRR security outcomes) [8]. Although the studies were limited in quantity, many of the elements, when applied to a community before a flood, were seen to reduce the harmful security and health outcomes after a severe flood. Similarly, the UK participants communicated their perceptions on how the nine elements were associated with the likely psychological impact, economic harm, and social disruption from a flood.

5.2 The Terminology of Community Resilience as It Applies to Disasters

In this book, three general interpretations of community resilience were found. However, these interpretations were blended in some areas of the literature and the discussions with focus group participants. 'Process' definitions, such as those given by Castleden et al. [10], Lemyre et al. [24], and Norris et al. [30], describe community resilience as an ongoing process or capability of change and adaptation in a community affected by a disaster. This change can happen before, during, and/or after a disaster. Similarly, many of the focus groups discussed community resilience as a process. Still, they focused on the way communities change after a disaster as prompted by the discussion of the nine elements. They saw the 'process' definitions of community resilience as more practical and consistent with their experiences of flooding than the 'bouncing back' narrative of community resilience. One participant summed this up nicely:

> Personally, the bouncing back is just wishful thinking. It's a much slower process, and I would say it's just a false way of projecting it because people get on with their lives as best as they can, but I wouldn't call it bouncing back quickly at all. [FG9]

One group expressed the perception that experiencing a disaster takes time to recover from and may not happen overnight [FG13]. Much of the literature and many focus group participants agreed that community resilience implied that a disaster-affected community must go through a recovery process to start functioning positively again after a disaster.

'Absence of adverse effect' definitions focused on the ability of a community to maintain stable functioning after a disaster. Papers that fit this type, such as Bonanno [6] and Gibson [19], viewed community resilience as an outcome after a disaster synonymous with positive functioning. Many focus group participants saw this type of definition of community resilience as 'bouncing back' after a disaster. They felt that they had no choice but to bounce back after a disaster [FG10] and return their community to positive functioning. They felt the pressure to resume back to life despite the security and health harm they faced.

Community resilience was also perceived as a broad collection of recovery and response-related abilities within a community. This definition was categorized as the 'range of attributes' definitions, as seen in the UK Cabinet Office report on community resilience [7] and the article by Coles and Buckle [16]. This definition from the literature matched with the "planning, protocols, and support" responses from many focus group participants. These responses included external support to having sufficient emergency measures in place within the community. Outcomes from the literature and focus groups were clear that if community resilience was defined in this way, it would include an interplay of abilities from within the community and post-disaster activities carried out by those from outside the community. The study participants found not enough support from outside the community to help them through the negative impacts caused by severe flooding.

A blending of the types of definitions was also identified when examining the term community resilience. For example, Pfefferbaum et al. [35, pp. 241–242] defined resilience as "…an attribute (e.g., ability, capacity), a process, and/or an outcome associated with successful adaptation to, and recovery from adversity", and that it "…differs depending on context and purpose". Highlighted by Pfefferbaum et al. [35] was the notion that community resilience was not only ambiguous and amorphous but also changeable depending on the context. This notion was echoed in the focus group discussions. One participant saw community resilience as a bouncing back accompanied by internal change and better awareness of flooding [FG13].

Many participants referred to a "range of attributes" definition when discussing that community resilience needs to capture a community's knowledge and the lessons identified and outcomes for action from its experience.

5.2.1 Summary

The meaning of community resilience was found to be ambiguous. Broad definitions from the literature and flood-affected participants were given and ranged from a process to response and recovery related definitions to a list of qualities needed for a community to reduce the damage caused by a flood. Future uses of community resilience should focus on the specificity of the nine constituent elements of community resilience commonly found in the literature.

5.3 The Elements of Community Resilience as They Apply to Severe Flooding

The term community resilience encompasses important elements that can help reduce the likely impacts of disasters on communities. Nine general elements were identified in a systematic review of the literature and were discussed in this book. Table 5.2 shows these elements and sub-elements. To see how these elements of community resilience are evident in the case of a specific disaster, flooding was used as the case example. Few studies were found to have clear associations between any elements of community resilience and the security and health harm following flooding. One exception was the multiple associations that were found between psychological impact and many of the elements.

The focus groups and survey results were able to provide further detail on how these applied to flooding, particularly in the UK. Below are short summaries of what was found for each element and its possible implications for the UK communities.

Local knowledge provided the focus group participants with confidence as they knew who to call or ask in case of a flood. The flood-affected participants were empowered through *local knowledge* to get involved and remain engaged in their community preparedness after the flood. This level of empowerment and engagement by the focus group participants is related to what Norris et al. [30] and Lowe et al. [28] see as the benefit of strengthening the competence and collective efficacy of a community against a future disaster. Even half of the survey participants believed their community would have enough local knowledge to cope with a flood in the first few days. A possible implication is building empowerment and engagement within the communities about flooding. For example, an awareness of the history of floods in the community and using an annual remembrance event could engage and empower community members around understanding their flood risk.

Community networks and relationships were easily understood and expressed by the flood-affected participants. Community members helping each other was one of

Table 5.2 Details on the elements of community resilience found in this book

Community resilience elements	Sub-elements	Key topics from focus groups
Local knowledge	Factual knowledge base	Taking actions on prior knowledge
		Knowing who to call or ask
		Community's history of rivers and floods
		Community awareness of flood risk
	Collective efficacy and empowerment	Community members buying or building their own resources
		Community members helping each other
	Training and education	Dangers of flooding, floodwater, and sewage
Community networks and relationships	Connectedness	Community connectedness
		Outside organization helping the community
	Cohesion	Community spirit
		Isolation and abandonment
Communication	Risk communication	Barriers to receiving communication
		Information from outside authorities
		Information from within the community
		Mistrust of communication
	Crisis communication	Barriers to receiving communication
		Information from outside authorities
		Information from within the community
		Mistrust of communication
	Effective communication	Effective communication
		Impact of media, filming crews, and tourists
Health	Health services	Quality and delivery of health services
	Physical health	Issues with physical health
	Mental health	Problems with mental health

(continued)

Table 5.2 (continued)

Community resilience elements	Sub-elements	Key topics from focus groups
Governance and leadership	Infrastructure and services	Sewage and clean water concerns
		Tankers and workers
		Action to alleviate sewage issues
		Having better leaders
		'Blame game' politics and lack of responsibilities from services and officials
	Public involvement and support	Community members giving crisis information to help with services
		Services and officials not being prepared or able to help communities
		Lack of public involvement and support
Resources	Resources	Wrong type or lack of resources
		Resources and essential goods bought by communities
		Leftover resources and post-flood improvements
		Resources that were said to help Sandbags
Economic Investment	Pre- and during disaster economic plan	Spending to prepare for a flood
		Spending during the flood
	Post-disaster economic programming	Government grants and post-disaster economic program
		Insurance companies and the strain of insurance
	Post-disaster economic development	Costs after the flood
		Post-disaster spending
Preparedness	Planning and mitigation	Lack of preparedness and plans
		Grassroots efforts by the communities
		Physical flood defenses
		Lessons identified and outcomes for action
		Future needs to improve communities' preparedness and crisis management

(continued)

Table 5.2 (continued)

Community resilience elements	Sub-elements	Key topics from focus groups
Mental outlook	Hope	Feelings of worried sick and not having any trust or faith
		Belief in never being able to recover and not having any hope for the future
		Complacency for things to change after the flood
		Outcome of long-lasting fear
	Adaptability	Adapting to a post-flood community
		Positive and negative coping experiences
		Previous flood experiences
		Sudden shock of the flood
		Trauma caused by the flood

the essential features described by the focus group participants that aided recovery. Rodriguez-Llanes et al. [39] and Aldrich and Meyer [5] suggest this kind of social support is beneficial and can be an indicator of a community that can endure a disaster. Even though most of the survey participants felt that people in their community helped each other, 48% of the survey participants disagreed that people in their community shared lessons on coping with flooding. A possible implication would be for practitioners to actively engage with community members about lessons identified from the previous flood and share strategies to cope with future floods. For example, holding public forums and community meetings to share lessons learned and coping strategies with past floods could increase community cohesion and connectedness to other communities with similar flood risk.

Receiving poor risk and crisis *communication* was a massive barrier to preventing and recovering from the flood impacts. Doubting the authenticity and communication being received within the community were also expressed by the focus groups. Participants mentioned this lack of trust in communication as they felt they could not believe the leadership of officials providing the information. From the survey results, a higher percentage of flood-risk participants agreed that they trusted community members and the accuracy of local media compared to public officials (46 vs. 28%). Increasing the likelihood that communication-based trust is built is a key implication for flood-risk communities in the UK. Building strong relationships among neighbors will hopefully reduce suspicion and increase trust in the community [5, 18, 30]. Transparency and accuracy in communication can also increase trust in messages and can lead communities to trust leaders [17, 36].

The *health* of the community, including the physical wellness and health services being provided, was a concern for many focus group participants, especially in terms of the trauma caused by the floods. Multiple flood-affected participants in Chap. 2 reported that the health services serving the community were overwhelmed, but once they arrived in the community, the help given was of high quality. Most of the survey participants (56%) agreed about the quality, but whether the health services available in their community would be quick to react to any disaster was not in clear agreement: 37% were unsure, and 41% agreed. For future directions, any possible delays in the delivery of health services to UK flood-risk communities should be examined and reduced even before a flood occurs. Simulations could help better prepare health services to face and overcome likely delays in servicing flood-prone areas. Building better practices will go a long way to providing better care in the community during and after a flood.

According to the focus group participants, there was a lack of *leadership and governance* seen in public services and infrastructure during their flood experience. Many of these flood-affected participants voiced their concerns that their prior interactions with government officials did not instill confidence or trust in their leadership, which has some overlap with overall poor communication. Many strongly believed that a future flood would likely be no different in receiving poor leadership from government officials. Over half of the survey participants (53%) were also unsure if their leaders would be effective in a disaster. An implication of these results is to prepare current leaders within and outside of the flood-risk communities with informed knowledge and better flood and disaster management skills. Incorporating more public involvement with the governance of flood risk management could help with enabling leaders to be better informed, such as providing a forum to listen to community members' experiences and concerns about the flood risk and crisis management systems. At the same time, a mix of formal and informal processes in communities help to reduce risk and build resilience [13, 14, 33]. Enabling more community members to be engaged with local and regional flood management systems and their decisions will also help build trust between the leaders and community members. Still, would community members even want to be involved? According to the focus group participants, the answer is yes. In a study by Adger et al. [3], the answer was yes as well if people perceived that everyone in the community received the same level of flood protection and if their concerns could be voiced and be listened to by the authorities in charge of the flood risk management decisions. For the element of governance and leadership, an emphasis should also be placed on re-establishing confidence and a better working relationship between the leaders and the community members before a flood occurs.

Most *resources*, besides sandbags, were said by the focus group participants to be provided by the community members rather than from any government authority or emergency service. These flood-affected participants discussed that many of their necessary daily resources and aid were from neighbors making runs through flooded streets for fellow flood-affected neighbors. Creating a formalized system where local supermarkets and other organizations help with food donations and deliver necessary resources for daily living should be discussed in areas of high flood risk. This type of formalized system could help reduce the outcome of survey participants being more unsure (44%) than agreeing (38%) that aid and help would be distributed justly in their community if a disaster had occurred. Having resources widely available and distributed fairly in the community are essential features for a community to reduce likely disaster impact [9, 43 and 44]. Better installation of distributive justice in flood-risk communities, such as having facilitated discussions on the matter among community members and having agreed on plans on resource allocation before a flood, could shift the uncertainty of the survey participants to ensuring fair, distributive justice will occur after a flood. The UK Government and aid organizations should focus on providing *resources* to build long-term systemic and sustainable solutions, such as making better infrastructure in the community to face a flood, rather than solely providing resources items like sandbags. Many focus group participants debated the effectiveness of sandbags.

The element of *economic investment* seemed to be frustrating for many focus group participants. Many participants were upset with the reimbursement process of the insurance companies and the complicated process of claiming government grants. More survey participants were unsure (46%) than in agreement (24%) about whether their community had the financial resources needed to deal with problems if their community were to be affected by a disaster. Future work is necessary to understand how such an investment in UK communities at risk from flooding can be effective, efficient, and sustained. For example, one study could build on the works by Lamond [22] and Lamond and Proverbs [23] to examine the specific cost–benefit of the type of investments within and outside of insurance that would yield positive effects against flooding in UK communities over time.

Furthermore, based on a 2016 report by the UK Enivornment Agency (EA) [12] on the costs and impacts of the 2013 and 2014 floods, close to £1,300 million ($1,700 million) was estimated as the damage costs in England and Wales. Of which, most financial and social costs were said to be felt by the residential community members. The EA report [12] goes on to call for collecting better data to improve the reliability of their calculations on the costs and impacts of floods, such as collecting primary data and ones from geographic information systems. The report's authors acknowledge the difficulty in knowing such data and learning precisely how the regional impacts of flooding affect the nation. These points brought up by the EA report are worth further investigation, especially examining how the local financial impact caused by flooding affects the economy at a national level. This type of investigation could lead to an immediate action of better investments in local communities because of the significant association with the national economy. A different kind of study that could lead to the direct action of assets into flood-risk communities would be carrying out and following a similar investigation as done by Aerts et al. [4]. Aerts et al. [4] calculated the cost estimates on strategies against flooding in New York City, which could be replicated and focused on the flood-prone and high-risk communities in the UK. Knowing these costs could enable policymakers and local leaders to understand and act with better cost-effective strategies against flooding in the UK.

The *preparedness* activities mentioned by the focus groups revolved around learning lessons and making sure things were suitable for the next flood. However, most survey participants either disagreed (40%) or were unsure (37%) if their community actively prepares for future disasters. Only 16% of these flood-risk participants confidently knew where their local disaster center would be if a disaster occurred. Finding new ways to influence a community to adopt a more proactive approach to disaster readiness is difficult and an ongoing challenge, but setting up education-based disaster preparedness programs can begin to help reduce the apparent lack of preparedness discussed by focus group participants and supported by the survey participants.

The participants' views and feelings on *mental outlook* as it applies to flooding were powerful. Despite being missed from widely cited reviews, models, and measurements related to community resilience (Table 5.3). The effects of the trauma caused by the flood and the uncertainty of the future were reported to be overwhelming and scary by many groups. However, the *mental outlook* is not defined as the mental health status of the community but rather a growth mindset enabling the community members to learn from previous incidents and endure and recover from future ones. This ties in with the ongoing distinction between definitions of resilience that are seen more as positive mental health status [6, 42], and posttraumatic growth [21, 25]. Both concepts are related to individual resilience more than community resilience but could contribute to the knowledge attained and retained from a flooding experience. Past reviews have conceptualized community resilience as being distinct from individual resilience or composition of people with strong individual resilience [17, 30, 34, 41]. However, the *mental outlook* could be the connecting idea that ties individual and community resilience together. Especially with further research and development of psychological resilience and posttraumatic growth, the interactions at individual and community levels for *mental outlook* could inform about the protection capacity against future harm for a community member.

5.3.1 Implications of Understanding the Elements of Community Resilience

Flood-affected participants in the focus groups stressed that they would like to see more actions to strengthen the nine elements of community resilience identified. Almost all groups confirmed that strengthening the elements would enhance their community and positively affect their lives when dealing with future flooding. What was found as lacking by the participants was the enhancement of these elements throughout the disaster cycle (pre-, during, and post-disaster). Examining these during a cycle can provide a deeper understanding of the elements' effectiveness and an evaluation of how programs and interventions reduce likely disaster impact.

A prospective cohort study would be an ideal next step to assess the effectiveness of the elements of community resilience in communities before and after an actual flood. Evidence could be collected on the specific effect that these elements have on the NRR security outcomes and whether the elements may differ across communities with varying risk levels, cultures, and flood exposure. This type of study could also examine the interconnectedness of community resilience and other disaster risk

Table 5.3 Elements of community resilience found in this book, compared to widely cited reviews, models and measurements

Factors from flood-risk participants	Theme/sub-theme from researchers and flood- affected participants	Resilience review [10]	CDR review [31]	RAND report [11]	DROP model [17]	Community resilience Index [30, 40]	CCRAM [15, 26]	CART [37, 38]
1. **Strong Fellowship** 2. **Trust in the System** 3. **Memories of Disasters**	**Local knowledge**	X	X	X	X	X		X
	Factual knowledge base		X		X	X		
	Collective efficacy and empowerment			X		X	X	X
	Training and education	X		X	X	X		
	Community networks and relationships	X	X	X	X	X	X	X
	Connectedness			X	X	X		X
	Cohesion	X		X	X	X		X
	Communication	X		X	X	X		X
	Risk communication			X	X			
	Crisis communication			X				
	Effective communication			X		X		
	Health			X	X	X		
	Health services			X		X		

(continued)

Table 5.3 (continued)

Factors from flood-risk participants	Theme/sub-theme from researchers and flood- affected participants	Resilience review [10]	CDR review [31]	RAND report [11]	DROP model [17]	Community resilience Index [30, 40]	CCRAM [15, 26]	CART [37, 38]
	Physical health	X		X				
	Mental health	X		X	X	X		X
	Governance and leadership	X	X	X	X	X	X	X
	Infrastructure and services		X		X	X		
	Public involvement and support			X	X			X
	Resources			X		X		X
	Economic investment	X	X	X	X	X		X
	Pre- and during disaster economic plan	X	X	X	X	X		X
	Post-disaster economic development		X	X	X	X		
	Post-disaster economic programming		X					
	Preparedness	X		X	X	X	X	X

(continued)

Table 5.3 (continued)

Factors from flood-risk participants	Theme/sub-theme from researchers and flood- affected participants	Resilience review [10]	CDR review [31]	RAND report [11]	DROP model [17]	Community resilience Index [30, 40]	CCRAM [15, 26]	CART [37, 38]
	Planning and mitigation	X		X	X	X		X
	Mental Outlook							
	Hope					X		
	Adaptability	X		X	X	X	X	X

reduction theories, such as the disaster vulnerability theory by Zakour and Gillespie [44] and the specific impact of reducing a community's vulnerability to a disaster beforehand [1, 2]. Reducing the likely consequences is hypothesized as what the elements could help if strengthened in a community before a disaster.

Another possible study could measure the elements of community resilience in a community before a disaster and check each element's contribution to reducing the community's vulnerability to a disaster. During and after, the elements can be measured against critical goals, such as those given by Norris et al. [30] and Chandra et al. [11] to make a community stronger after a disaster. These goals after a disaster were to enhance a community's recovery, inform about new vulnerabilities, and strengthen the community's preparedness for the next disaster. Assessing each element against each goal could tell whether each element has a presence, to what degree the elements are proving to be effective in the community, and to what effect the elements are having on the NRR security outcomes during and after a disaster. Finally, due to the exploratory nature of the factor analysis on the cross-sectional survey (Chap. 3), clarification will also be needed on whether the elements of community resilience overlap and, if so, to what degree. The elements of community resilience were found in an exploratory factor analysis to blend into three factors expressing general ideas of fellowship, trust, and memories in a community. However, the flood-affected participants from Chap. 2 identified the elements with specific examples of how aspects of the elements helped them reduce their flood impact, thus, why they were determined separately.

Table 5.3 presents how the elements compare to widely cited reviews, models, and measurements on the concept of community resilience. Only the model developed in the RAND report [11] and the Community Resilience Index [40] based on the paper by Norris et al. [30] are inclusive enough to address close to all the elements and sub-elements found in this book. Neither of the existing two scales developed prior to the qualitative and quantitative research conducted in this Book, the Communities Advancing Resilience Toolkit (CART) and the Conjoint Community Resiliency Assessment Measure (CCRAM), are comprehensive to include all the elements. Also, clarification is needed on the best method to measure the elements of community resilience as CART and CCRAM have different methods. For example, the CART [37, 38] provides a list of areas to build resilience in a community. The authors devised a toolkit that acts as a community-based participatory intervention that will develop capacities and attributes in community leaders and members that should help to build resilience against a disaster.

On the other hand, CCRAM [15, 26] uses a cross-sectional survey to determine resilience based on preparedness, leadership, place attachment, social trust, and collective efficacy. Based on their recent publication on their pilot study of monitoring communities before and after an emergency using their measurement tool, Leykin et al. [27] saw that their elements of community resilience could grow weaker or stronger over time. This finding should be followed up with more studies to confirm this aspect and see the elements' specific ebb and flow mechanisms throughout the disaster cycle. Additionally, future studies should also look at how the interconnectedness of the elements affects the communities against likely and actual disaster impacts.

An ideal starting point for flood-specific disasters would be to integrate all of the nine elements of community resilience with the existing models and interventions used to strengthen communities against flood risk. Ensuring all the elements are present and monitored in the community will develop the concept of community resilience as it applies to flooding further. As reported in the findings in Chap. 2, even participants who were flooded in the same community did not always see or express community resilience in the same way. This brings doubt over whether a program or intervention calling for "better community resilience" would be readily understood by the desired target audience.

5.3.2 *Summary*

Flood-affected participants and high flood-risk participants were asked about the elements of community resilience and their relation to reducing disaster impact. All of the elements of community resilience were discussed, and each application to disaster impact was described with specific details and examples. These ranged from how an element helped the participants reduce their flood impact to how participants lacking in an element resulted in a perceived negative impact. One of the main achievements was creating a list of specific details on which aspects flood-affected participants found significant among the elements of community resilience. Practitioners and researchers should use this list to guide what areas to examine and discuss while working in flood-affected and flood-risk communities in the UK.

5.4 Limitations of This Book

Although limitations were highlighted within each chapter, it is essential to consider the broader restrictions that may apply to the entire book. First, the author is not a native to the UK as a researcher and practitioner. This limits the exposure to and understanding of possible socio-political and cultural factors that may affect community resilience to flooding in the UK. However, the author found that the focus group participants were comfortable explaining these factors and their impact in detail due to being an outsider. This allowed more information to be obtained from them, as they assumed that the author did not know anything about it.

Second, the scope of work must be married with the assumptions made with the terminology relating to a disaster, flood, community, and locality used throughout the chapters. For example, a specific definition for a disaster given by UNISDR was used, while the term 'community' was broadly considered a place with a population of people. The methods for Chap. 4 assumed the meaning of a flood to be any type of natural hazard that brought an abundance of water into a community. Logically, the disaster type will alter the prominence of the elements of community resilience as it applies to likely harm caused by a disaster. However, this book broadly assumed that the shifting of emphasis that may have occurred by the different terminologies used between chapters was minimal when distinguishing the possible elements. For example, in Chap. 2, participants were defined as having been flooded by self-reporting. As a result, the discussions included people who had experienced varying degrees of impacts caused by the river, coastal, surface water, sewer, and groundwater flooding. The participants had different understandings of what it meant to be flooded—from water entering their homes to water being a disruption outside their homes. Also, the focus group participants considered their community to be their fellow neighbors and their locality to be the greater local authority. According to the EA flood risk maps, flood risk was determined by postcodes in survey participants in Chap. 3. Flood-risk participants provided their own interpretation of the community before the survey. After which, the survey provided a broad meaning of community to the participants to use for the survey questions. This given definition of community assumed a population-based concept but was different from the one used in the comprehensive review by Patel et al. [32] and Chap. 2. These differing interpretations of what constitutes a community could have limited the understanding of the concept of community resilience. It was assumed that community resilience could be homogeneously applied to all types of communities for various disaster types.

Third, the literature reviews used in this book were limited by location, disaster type, and quality. For example, in Patel et al. [32], the studies were found in different cultures and locations, such as America, Israel, and Australia, along with varying types of disasters such as hurricanes, earthquakes, and bushfires. Chapter 4 included seven publications from the USA, six from the UK, two from Poland, two from Australia, and one from China and Mexico when looking at flooding specifically. Within these 19 publications, the quality and methodology of the research varied, with most studies being poor indicators of whether the elements of community resilience were associated with NRR security outcomes. Having better quality research focusing on the same type of disaster within the same culture type could be important for understanding community resilience.

Fourth, the resources available for this research were limited. After carrying out the focus groups, more funding and resources could have enabled and established a cohort study. Instead, available funding covered the costs of a cross-sectional study.

Lastly, when discussing community resilience as it applies to flooding, most of the focus group participants in this book may have been thinking of community resilience retrospectively, which for most cases was the 2013/14 flood in the UK. Participants of the cross-sectional survey may also have been answering using their last memory or observation from the media of a past flood in the UK, which again was likely the 2013/14 flood. The findings from these participants may only apply to the context of this incident and may not be generalizable to other floods. Further research will have to confirm how the results of this book apply to other floods. This limitation would be the same if using the findings for other non-flood disasters.

5.5 Conclusions

Communicating community resilience is difficult when the concept is interpreted and presented ambiguously. However, this book found consistencies in the way practitioners and researchers, flood-affected people, and people at risk of flooding understand community resilience broadly. From reviewing existing definitions of community resilience, no over-arching definition was found, but nine common themes were identified for how communities can be resilient against disasters. Using the context of flooding in the UK, the themes associated with community resilience found by researchers were elaborated on by flood-affected participants and those in settings deemed at risk for flooding. Based on reviews of the scientific literature, discussions from focus groups and interviews, and results from a cross-sectional survey, the learnings presented in this book present novel research outcomes and areas for future work on applying community resilience (Table 5.4).

There is much to be understood about community resilience as it applies to disasters, such as the complexities involved in using this concept before, during, and after a disaster. Although there are differing views on what community resilience is, the findings of this book acknowledge the concept of community resilience to be essential for managing disasters that cause reduced environmental security. Further understanding of what the concept entails may enable future UK Government and UN initiatives to be more comprehensive in their coverage of this crucial topic. Future directions involve discussing the nine core elements explicitly with practitioners and researchers engaged with community resilience, local knowledge; community networks and relationships; communication; health, governance, and leadership; resources; economic investment; preparedness; and mental outlook. They should also be used in discussion with flood-affected people around this concept. For researchers and practitioners, reducing the emphasis on encompassing definitions of community resilience and instead focusing on the constituent elements will be more helpful in understanding how to mitigate likely harm due to reduced environmental security in communities. When applying it to practice, themes exist across groups and have the potential to reduce likely security and health harm caused by severe flooding in communities.

Table 5.4 Summary of outcomes of this book, highlighting the novel research outcomes and areas for future work

	What was found	What needs caution and further research
Likely Application of Community Resilience	**Concept:**	**Concept:**
	• Several models and definitions of Community Resilience exist, and the elements used in this book encompass them all (as seen in Table 5.3) • Overlooking the ambiguous interpretations, exploring the commonalities captured by the nine elements were the conclusions found and supported by literature, focus groups, and survey	• Further research on definitions may be unnecessary as the existing definitions contrasted too much, and future discussions could simply be a distraction • Exploring and applying a model that encompasses pre-, during, and post-disaster settings and how the elements overlap and work would advance conceptual discussions further
	Measurement:	**Measurement:**
	• More measurement tools have been published since compiling this book, but each tool differed in its definition of community resilience • Currently, a comprehensive tool to measure community resilience does not exist, as the term is not well-defined, nor is there agreement that community resilience is a successfully measured outcome	• In the future, researchers need to use caution when applying the term community resilience to their measurement tools • A study to examine the elements of community resilience as they apply to flooding with their association to NRR outcomes should be carried out in the UK
	Practice:	**Practice:**
	• Providing activities related to community resilience elements will help reduce harm in flood-affected communities • Activities framed and applied in communities to strengthen the fellowship and trust in the community and learn from past disasters were understood by the flood-risk public as critical elements of community resilience	• More effort needs to be made on developing preparedness activities related to the elements of community resilience with communities before a disaster occurs • Further studies should monitor and evaluate activities related to the elements of community resilience throughout the disaster cycle and distinguish the effectiveness and efficacy of each element in different disaster settings

References

1. Adger WN (2006) Vulnerability. Glob Environ Chang 16(3):268–281. https://doi.org/10.1016/j.gloenvcha.2006.02.006
2. Adger WN, Hughes TP, Folke C, Carpenter SR, Rockstrom J (2005) Social-ecological resilience to coastal disasters. Science 309(5737):1036–1039. https://doi.org/10.1126/science.1112122
3. Adger WN, Quinn T, Lorenzoni I, Murphy C (2016) Sharing the pain: perceptions of fairness affect private and public response to hazards. Ann Am Assoc Geogr 106(5):1079–1096. https://doi.org/10.1080/24694452.2016.1182005
4. Aerts JC, Botzen WJ, de Moel H, Bowman M (2013) Cost estimates for flood resilience and protection strategies in New York City. Ann N Y Acad Sci 1294:1–104. https://doi.org/10.1111/nyas.12200
5. Aldrich DP, Meyer MA (2015) Social capital and community resilience. Am Behav Sci 59(2):254–269. https://doi.org/10.1177/0002764214550299
6. Bonanno GA (2004) Loss, trauma, and human resilience: have we underestimated the human capacity to thrive after extremely aversive events? Am Psychol 59(1):20–28. https://doi.org/10.1037/0003-066x.59.1.20
7. Cabinet Office (2010) Draft strategic national framework on community resilience. London, UK. Retrieved from https://www.gov.uk/government/uploads/system/uploads/attachment_data/file/60922/Strategic-National-Framework-on-Community-Resilience_0.pdf
8. Cabinet Office. (2015). National risk register for civil emergencies-2015 edition. London, UK. https://www.gov.uk/government/uploads/system/uploads/attachment_data/file/419549/20150331_2015-NRR-WA_Final.pdf
9. Carlson J, Haffenden R, Bassett G, Buehring W, Collins M III, Folga S, et al (2012) Resilience: theory and application. U.S. Department of Energy, Argonne, IL. Retrieved from http://www.ipd.anl.gov/anlpubs/2012/02/72218.pdf
10. Castleden M, McKee M, Murray V, Leonardi G (2011) Resilience thinking in health protection. J Public Health 33(3):369–377. https://doi.org/10.1093/pubmed/fdr027
11. Chandra A, Acosta J, Stern S, Uscher-Pines L, Williams M, Yeung D, et al (2011) Building community resilience to disasters: a way forward to enhance national health security (technical report). RAND Corporation, Arlington, VA. Retrieved from https://www.rand.org/content/dam/rand/pubs/technical_reports/2011/RAND_TR915.pdf
12. Chatterton J, Clarke C, Daly E, Dawks S, Elding C, Fenn T, et al (2016) The costs and impacts of the winter 2013 and 2014 floods. (SC140025/R1). Environment Agency, Bristol, UK. Retrieved from https://www.gov.uk/government/uploads/system/uploads/attachment_data/file/501784/The_costs_and_impacts_of_the_winter_2013_to_2014_floods_-_report.pdf
13. Clark-Ginsberg A, Blake JS, Patel KV (2022) Hybrid governance and disaster management in Freetown, Sierra Leone, Monrovia, Liberia, and Dar es Salaam Tanzania. Disasters 46(2):450–472. https://doi.org/10.1111/disa.12466
14. Clark-Ginsberg A, Easton-Calabria LC, Patel SS, Balagna J, Payne LA (2021) When disaster management agencies create disaster risk: a case study of the US's Federal Emergency Management Agency. Disaster Prev Manag: Int J. Chicago
15. Cohen O, Leykin D, Lahad M, Goldberg A, Aharonson-Daniel L (2013) The conjoint community resiliency assessment measure as a baseline for profiling and predicting community resilience for emergencies. Technol Forecast Soc Chang 80(9):1732–1741. https://doi.org/10.1016/j.techfore.2012.12.009
16. Coles E, Buckle P (2004) Developing community resilience as a foundation for effective disaster recovery. Austr J Emergency Manag 19(4):6. Retrieved from https://search.informit.com.au/documentSummary;dn=375435145094637;res=ielhss
17. Cutter SL, Barnes L, Berry M, Burton C, Evans E, Tate E et al (2008) A place-based model for understanding community resilience to natural disasters. Glob Environ Chang 18(4):598–606. https://doi.org/10.1016/j.gloenvcha.2008.07.013

18. Dawes SS, Cresswell AM, Cahan BB (2004) Learning from crisis: lessons in human and information infrastructure from the world trade center response. Soc Sci Comput Rev 22(1):52–66. https://doi.org/10.1177/0894439303259887
19. Gibson CA (2010) An integrated approach to managing disruption-related risk: Life and death in a model community. J Bus Cont Emergency Plan 4(3):246–261. Retrieved from http://www.ingentaconnect.com/content/hsp/jbcep/2010/00000004/00000003/art00006
20. Hanlon HM, Bernie D, Carigi G, Lowe JA (2021) Future changes to high impact weather in the UK. Clim Change 166(3):50. https://doi.org/10.1007/s10584-021-03100-5
21. Hobfoll SE, Hall BJ, Canetti-Nisim D, Galea S, Johnson RJ, Palmieri PA (2007) Refining our understanding of traumatic growth in the face of terrorism: moving from meaning cognitions to doing what is meaningful. Appl Psychol 56(3):345–366. https://doi.org/10.1111/j.1464-0597.2007.00292.x
22. Lamond J (2011) Financial implications of flooding and the risk of flooding on households. In: Lamond J, Booth C, Hammond F, Proverbs D (eds) Flood hazards: impacts and responses for the built environment. CRC Press, pp 317–326. https://doi.org/10.1201/b11050-28
23. Lamond J, Proverbs D (2006) Does the price impact of flooding fade away? Struct Surv 24(5):363–377. https://doi.org/10.1108/02630800610711960
24. Lemyre L, Clément M, Corneil W, Craig L, Boutette P, Tyshenko M et al (2005) A psychosocial risk assessment and management framework to enhance response to CBRN terrorism threats and attacks. Biosecur Bioterror 3(4):316–330. https://doi.org/10.1089/bsp.2005.3.316
25. Levine SZ, Laufer A, Stein E, Hamama-Raz Y, Solomon Z (2009) Examining the relationship between resilience and posttraumatic growth. J Trauma Stress 22(4):282–286. https://doi.org/10.1002/jts.20409
26. Leykin D, Lahad M, Cohen O, Goldberg A, Aharonson-Daniel L (2013) Conjoint community resiliency assessment measure-28/10 items (CCRAM28 and CCRAM10): a self-report tool for assessing community resilience. Am J Commun Psychol 52(3–4):313–323. https://doi.org/10.1007/s10464-013-9596-0
27. Leykin D, Lahad M, Cohen R, Goldberg A, Aharonson-Daniel L (2016) The dynamics of community resilience between routine and emergency situations. Int J Disaster Risk Reduct 15:125–131. https://doi.org/10.1016/j.ijdrr.2016.01.008
28. Lowe SR, Joshi S, Pietrzak RH, Galea S, Cerda M (2015) Mental health and general wellness in the aftermath of Hurricane Ike. Soc Sci Med 124:162–170. https://doi.org/10.1016/j.socscimed.2014.11.032
29. Moore M, Chandra A, Feeney KC (2012) Building community resilience: what can the United States learn from experiences in other countries? Disaster Med Public Health Prep. https://doi.org/10.1001/dmp.2012.15
30. Norris FH, Stevens SP, Pfefferbaum B, Wyche KF, Pfefferbaum RL (2008) Community resilience as a metaphor, theory, set of capacities, and strategy for disaster readiness. Am J Commun Psychol 41(1–2):127–150. https://doi.org/10.1007/s10464-007-9156-6
31. Ostadtaghizadeh A, Ardalan A, Paton D, Jabbari H, Khankeh H (2015) Community disaster resilience: a systematic review on assessment models and tools. PLOS Curr Disast. https://doi.org/10.1371/currents.dis.f224ef8efbdfcf1d508dd0de4d8210ed
32. Patel SS, Rogers MB, Amlôt R, Rubin GJ (2017) What do we mean by 'community resilience'? A systematic literature review of how it is defined in the literature. PLoS Curr 9
33. Patel SS, McCaul B, Cáceres G, Peters LE, Patel RB, Clark-Ginsberg A. (2021) Delivering the promise of the sendai framework for disaster risk reduction in fragile and conflict-affected contexts (FCAC): a case study of the NGO GOAL's response to the Syria conflict. Prog Disaster Sci 10:100172
34. Paton D (2008) Community resilience: integrating individual, community and societal perspectives. In: Gow K, Paton D (eds) The phoenix of natural disasters: community resilience. Nova Science, New York, pp 13–21. Retrieved from https://books.google.com/books?id=FEj7-FvB8xgC&dq=paton+community+resilience+2008&lr=&source=gbs_navlinks_s
35. Pfefferbaum B, Pfefferbaum RL, Van Horn RL (2015a) Community resilience interventions: participatory, assessment-based, action-oriented processes. Am Behav Sci 59(2):238–253. https://doi.org/10.1177/0002764214550298

36. Pfefferbaum RL, Pfefferbaum B, Nitiéma P, Houston JB, Van Horn RL (2015b) Assessing community resilience: an application of the expanded CART survey instrument with affiliated volunteer responders. Am Behav Sci 59(2):181–199. https://doi.org/10.1177/0002764214550295
37. Pfefferbaum RL, Pfefferbaum B, Van Horn RL (2011) Communities advancing resilience toolkit (CART): the CART integrated system. Oklahoma City, OK. Retrieved from https://www.oumedicine.com/docs/ad-psychiatry-workfiles/cart_online-final_042012.pdf?sfvrsn=2
38. Pfefferbaum RL, Pfefferbaum B, Van Horn RL, Klomp RW, Norris FH, Reissman DB (2013) The communities advancing resilience toolkit (CART): An intervention to build community resilience to disasters. J Public Health Manag Pract 19(3):250–258. https://doi.org/10.1097/PHH.0b013e318268aed8
39. Rodriguez-Llanes JM, Vos F, Guha-Sapir D (2013) Measuring psychological resilience to disasters: are evidence-based indicators an achievable goal? Environ Health Glob Access Sci Source 12:115. https://doi.org/10.1186/1476-069X-12-115
40. Sherrieb K, Norris F, Galea S (2010) Measuring capacities for community resilience. Soc Indic Res 99(2):227–247. https://doi.org/10.1007/s11205-010-9576-9
41. Walsh F (2007) Traumatic loss and major disasters: strengthening family and community resilience. Fam Process 46(2):207–227
42. Westphal M, Bonanno GA (2007) Posttraumatic growth and resilience to trauma: different sides of the same coin or different coins? Appl Psychol 56(3):417–427. https://doi.org/10.1111/j.1464-0597.2007.00298.x
43. White RK, Edwards WC, Farrar A, Plodinec M (2015) A practical approach to building resilience in America's communities. Am Behav Sci 59(2):200–219. https://doi.org/10.1177/0002764214550296
44. Zakour MJ, Gillespie DF (2013) Community disaster vulnerability: theory, research, and practice. Springer, New York, NY. Retrieved from https://www.springer.com/gp/book/9781461457367

MIX
Papier aus verantwortungsvollen Quellen
Paper from responsible sources
FSC® C105338

If you have any concerns about our products,
you can contact us on
ProductSafety@springernature.com

In case Publisher is established outside the EU,
the EU authorized representative is:
Springer Nature Customer Service Center GmbH
Europaplatz 3, 69115 Heidelberg, Germany

Printed by Libri Plureos GmbH
in Hamburg, Germany